Calculus
and
Pizza

ALSO BY CLIFFORD A. PICKOVER

Calculus and Pizza

A Cookbook for the Hungry Mind

CLIFFORD A. PICKOVER

WILEY

John Wiley & Sons, Inc.

This book is dedicated to people who love mathematics and pizza.

This book is also dedicated to those people who consume

$$\left(\int_0^1 x^2 dx \right)^{-1}$$

garlic pizzas every week.

Copyright © 2003 by Clifford A. Pickover. All rights reserved
Illustrations on pages xi, 2, 4, 5, 14, 30, 42, 46, 49, 53, 54, 83, 85, 86, 88, 89, 90, 96, 112, 120, 125, and 126 copyright © 2003 by B. K. Mansfield. All rights reserved

Published by John Wiley & Sons, Inc., Hoboken, New Jersey
Published simultaneously in Canada

Design and production by Navta Associates, Inc.

For general information about our other products and services, please contact our Customer Care Department within the United States at (800) 762-2974, outside the United States at (317) 572-3993 or fax (317) 572-4002.

Wiley also publishes its books in a variety of electronic formats. Some content that appears in print may not be available in electronic books. For additional information about Wiley products, visit our web site at www.wiley.com.

Library of Congress Cataloging-in-Publicaton Data

Pickover, Clifford A.
  Calculus and pizza : a cookbook for the hungry mind / Clifford A. Pickover.
    p. cm.
  Includes bibliographical references and index.
  ISBN 0-471-26987-5 (pbk.)
    1. Calculus. I. Title.
  QA203.2.P53 2003
  515—dc21   2003050192

Printed in the United States of America

10 9 8 7 6 5 4 3 2 1

Taking mathematics from the beginning of the world
to the time of Newton, what he has
done is much the better half.

—Gottfried Wilhelm Leibniz, codiscoverer of calculus

■ ■ ■

If I have seen further than others,
it is by standing upon the shoulders of giants.

—Isaac Newton, codiscoverer of calculus

■ ■ ■

If I have not seen as far as others,
it is because giants were standing on my shoulders.

—Hal Abelson, MIT professor

Contents

Preface

▀▟

> The invention of the calculus was one of the great intellectual achievements of the 1600s. By one of those curious coincidences of mathematical history not one, but two men devised the idea—and almost simultaneously.
>
> —David M. Burton, *The History of Mathematics*

> Non cooks think it's silly to invest two hours' work in two minutes' enjoyment; but if cooking is evanescent, well, so is the ballet.
>
> —Julia Child, chef, author, TV hostess

Why This Book?

One of the abiding sins of mathematicians is an obsession with completeness—an urge to go back to first principles to explain their work. As a result, readers must often wade through pages of background before getting to the essential ingredients. To avoid this burden, each chapter in this book is short. You'll quickly get the essence of a technique or question. A number of the basic methods and challenges are representative of a wider class of problems of interest to mathematicians today. One advantage of this brief format is that you can jump right in to experiment and have fun without having to sort through a lot of detritus. Thus, this book is not intended for mathematicians looking for formal mathematical explanations. Of course, this approach has some disadvantages. In just a few pages, our chef, Luigi, can't go into any depth on a subject. You won't find much historical context, philosophy, or extended discussions. You won't find proofs and derivations, though most of the important formulas are presented. The focus is on procedures and drills for solving problems, not on some deeper meaning of calculus.

I became interested in calculus early in childhood when gazing at a book in my father's dusty study. The book contained multiple integrals like

$\int \int \int$. These strange symbols impressed the heck out of me. I wondered if I would I ever be able to understand what an integral was. My father's books provided a seed from which my interest in calculus grew, and, along with my love of pizza, provided an early stimulus for *Calculus and Pizza*.

Calculus and Pizza is for anyone who has pondered what calculus is all about, for students who want to do well on exams and get into good colleges or graduate schools, and for lay people who want to get just a taste or a refresher course. The book might most effectively be used as a supplement to a traditional calculus text. If you haven't taken calculus for a few years, then *Calculus and Pizza* can serve as a quick review of some of the essential rules, formulas, and problems. I assume that the reader is familiar with high school algebra and some of the basic rules of trigonometry.

I hope that *Calculus and Pizza* will stimulate creative thinking, get some students interested in computer programming, and suggest the usefulness of simple mathematics for solving curious, practical, or even mindshattering problems. The ability of our future workforce to meet our needs for mathematicians, scientists, and engineers depends on the education that today's students receive. Recent statistics indicate that America is in grave danger of failing to meet the challenge. For much of the 1990s, U.S. students ranked low on scores in science and math in comparison to other countries. I hope *Calculus and Pizza* is the first book in a series that makes complex topics in science and math a bit more fun while emphasizing basic and practical principles. I welcome reader feedback and can be contacted through my web page at www.pickover.com.

A 60-Second History of Calculus

> And what are these fluxions [derivatives]? The velocities of evanescent increments. And what are these same evanescent increments? They are neither finite quantities, nor quantities infinitely small, nor yet nothing. May we not call them ghosts of departed quantities?
>
> —George Berkeley (1685–1753), a metaphysician who attacked Newton's foundations of calculus

Those of you who are bored by history may skip this section and proceed to the Introduction, but you'll miss a feud between mathematicians that rivals that of the Hatfields and the McCoys. Also, I should warn you that some of the terms used in this section, like *differentiation* and *integration,* won't make much sense to you until you finish this book—something you may well be looking forward to. For those of you who *are* curious about the origin of calculus, note that the English mathematician Isaac Newton (1642–1727) and the German mathematician Gottfried Wilhelm Leibniz (1646–1716) are generally credited with its invention (Figure 1), but various earlier mathematicians explored the concept of rates and limits, starting with the ancient Egyptians, who developed rules for calculating the volumes of pyramids and approximating the areas of circles. Greeks like Archimedes continued with more advanced studies of areas and volumes. More recent mathematicians, like René Descartes (1596–1650) and Pierre de Fermat (1601–1665), used techniques approaching those of modern calculus for finding slopes of tangents to curves and finding minima and maxima for functions. The English mathematicians John Wallis (1616–1703) and Isaac Barrow (1630–1677; Isaac Newton's teacher) also contributed to the development of calculus.

FIGURE 1. The mathematicians Isaac Newton (left) and Gottfried Wilhelm Leibniz (right) arguing about who should receive credit for the invention of calculus.

In the 1600s, both Newton and Leibniz puzzled over problems of tangents, rates of change, minima, maxima, and infinitesimals (unimaginably tiny quantities that are almost but not quite zero). Both men understood that differentiation (finding tangents to curves) and integration (finding areas under curves) are inverse processes. Newton's discovery (1665–1666) started with his interest in infinite sums; however, he was slow to publish his findings. Leibniz published his discoveries of differential calculus in 1684 and integral calculus in 1686. He said, "It is unworthy of excellent men, to lose hours like slaves in the labor of calculation. . . . My new calculus . . . offers truth by a kind of analysis and without any effort of imagination." Newton was outraged. Debates raged for many years on how to divide the credit for the discovery of calculus, and, as a result, progress in calculus was delayed. (More on this feud can be found in this book's Conclusion.) Today we use Leibniz's symbols in calculus such as $\frac{df}{dx}$ for the derivative and the $\int$ symbol for integration. (This integral symbol was actually a long letter S for *summa,* the Latin word for "sum.") The mathematician Joseph Louis Lagrange (1736–1813) was first to use the notation $f'(x)$ for the first derivative and $f''(x)$ for the second derivative. In 1696, Guillaume François de L'Hôpital, a French mathematician, published the first textbook on calculus.

The simultaneous discovery of calculus by Newton and Leibniz makes me wonder why so many discoveries in science were made at the same time by people working independently. For example, Charles Darwin (1809–1882) and Alfred Wallace (1823–1913) both developed the theory of evolution independently. In fact, in 1858, Darwin announced his theory in a paper presented at the same time as a paper by Wallace, a naturalist who had also developed the theory of natural selection. The mathematicians János Bolyai (1802–1860) and Nikolai Lobaschevcky (1793–1856) developed hyperbolic geometry independently at the same time. Most likely, such simultaneous discoveries have occurred because the time was ripe for them, given humanity's accumulated knowledge during a specific time frame. However, mystics have suggested that there is a deeper meaning to such coincidences. The Austrian biologist Paul Kammerer (1880–1926) wrote, "We thus arrive at the image of a world-mosaic or cosmic kaleidoscope, which, in spite of constant shufflings and rearrangements, also takes care of bringing like and like together." He compared events in our world to the tops of ocean waves, which seem isolated and unrelated. According to his controversial theory, we notice the tops of the waves, but beneath the surface there may be some kind of synchronistic mechanism that mysteriously connects events in our world and causes them to cluster. Whatever the case, since the time of Newton and Leibniz, calculus has made an indelible impact on science and society.

A Note on the Word "Calculus"

Before starting this book, I should note that the word *calculus* in medicine refers to an abnormal concretion of mineral salts occurring in a body organ such as the bladder, gallbladder, and pancreatic, salivary, and prostate glands. These calculi or stones have been known since ancient times. For example, the ancient Sumerians and Egyptians discovered kidney calculi. Today we know that such calculi consist of calcium and magnesium salts. Bladder calculi often assume enormous size, sometimes 11 centimeters in diameter. This book does not discuss the medical variety of calculus.

Acknowledgments

I thank Brian Mansfield for his wonderful cartoon diagrams used throughout this book. Over the years, Brian has been helpful beyond compare. I also owe a special debt of gratitude to many wonderful calculus books that have been published in the past, and these are cited in the Further Reading section. I heartily recommend these books for further information on calculus and its applications.

I thank Dennis Gordon, Nick Hobson, Les Axelrod, Robert Stong, and Ann Ostberg for their useful comments and suggestions.

Introduction

It's fun to get together and have something good to eat at least once a day. That's what human life is all about—enjoying things.

—Julia Child

Pizza is a lot like calculus. Except the limit is how much you can eat.

—Luigi

Calculus is the hammer that shatters the ice of our unconscious.

—Big Tony (Luigi's friend)

Hello. My name is Luigi. I own a pizza shop in New York City. If you are reading this book, you can probably guess my two main passions: calculus and pizza. I love pizza for its taste and its Neapolitan origin. I, too, am from Naples. I love calculus because it is an intellectual triumph and, like pizza, can be appreciated by devouring a small slice at a time.

Pizza Velocity and the Derivative

The calculus is the greatest aid we have to the appreciation of physical truth in the broadest sense of the word.

—William F. Osgood, *Mathematical Maxims and Minims*

The only real stumbling block is the fear of failure. In cooking, you've got to have a what-the-hell attitude.

—Julia Child

I throw my pizza dough high into the air. My arms are strong. My ceiling is high. Once the pizza is near the ceiling, it falls for many seconds before landing in my arms. I plan to add pepperoni to the pizza but am not concerned about it yet. It has nothing to do with this problem. It just tastes good.

Let's look at the pizza falling from near the ceiling. Suppose that the distance in feet that the pizza travels after t seconds is given by the formula

$$\text{feet (as a function of seconds)} = f(t) = 16t^2.$$

The notation $f(t)$ indicates that the distance is a function of time. After one-tenth of a second, the pizza has fallen $f(0.1) = 16(0.1)^2 = 0.16$ feet. After two-tenths of a second, the pizza has traveled 0.64 feet. After three-tenths of a second, the pizza has traveled 1.44 feet and so forth. The pizza is speeding up! I had better not miss catching the pizza; we don't want a mess.

How fast is the pizza traveling just as I catch it? Can we find a formula that describes the pizza's velocity at *every* instant? It's an intriguing question, but that's what my series of lessons is about. We can calculate the speed using calculus. Let me write this on the wall of my restaurant so that all my customers can see it (Figure 3).

B.C. MANSFIELD

FIGURE 3. Luigi ready to teach calculus.

Calculus seems mystical to me. The tricky thing about computing the velocity at an *instant* is that the distance traveled by the pizza and the elapsed time are both zero. That's what we mean by the word *instant*. It's as if we've carved up the fabric of reality and taken a slice to study. To find the velocity of the pizza at any instant, we can't just divide the distance traveled by the time it took the pizza to travel this distance. In fact, (total distance) ÷ (total time) would give us the *average velocity* of the moving pizza, but we know that the pizza's velocity is actually changing all the time. Instead, to determine the instantaneous velocity, we want to compute the average velocity over shorter and shorter time periods of length Δt. (The symbol Δ is the Greek letter *delta*. Mathematicians like to use it to symbolize a difference or a change. So, Δt is the difference between two instances in time. Δt might be 1 second or 0.001 second [a millisecond], or some other time interval.)

Let's try to calculate the distance traveled by the pizza between time t and $(t + \Delta t)$. The expression $(t + \Delta t)$ stands for a point in time that is just a short while later than the initial time t as indicated in Figure 4.

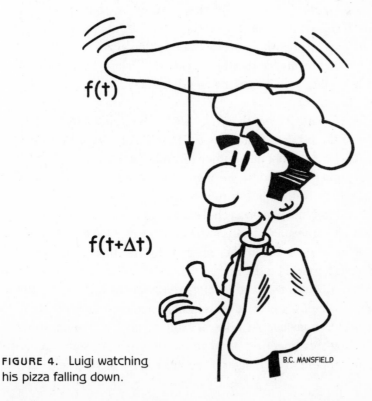

FIGURE 4. Luigi watching his pizza falling down.

Let's think about average velocity: (total distance) ÷ (total time). To obtain the distance traveled for a period of time, we need to subtract the distance the pizza traveled between the starting time t and the ending time $(t + \Delta t)$. The difference in distances can be written as $f(t + \Delta t) - f(t)$ because f describes the location of the falling pizza at any time. To find out the actual distances involved, we use our formula for distance: $f(t) = 16t^2$, to get $16(t + \Delta t)^2 - 16t^2$. If you recall the binomial formula, which says $(a + b)^2 = a^2 + 2ab + b^2$, you can quickly see that the $(t + \Delta t)^2$ is equal to $t^2 + 2t\Delta t + (\Delta t)^2$. Thus, the distance traveled by the pizza is $16(t^2 + 2t\Delta t + \Delta t^2) - 16t^2$. This is the same as $16(t^2 + 2t\Delta t + \Delta t^2 - t^2)$, and the t^2 terms cancel each other out. Thus, we know that the distance traveled by the pizza between time t and a while later $(t + \Delta t)$ is $16(2t\Delta t + \Delta t^2)$.

Are you with me so far? If not, review the algebra in the paragraph again and again until the manipulation becomes second nature to you.

We now have the distance traveled by the pizza during a time interval. If we divide this distance $16(2t\Delta t + \Delta t^2)$ by Δt, which is the time needed to travel the distance, we see that the average velocity over this period of time is $16(2t + \Delta t)$. Now for the fun part. As Δt gets smaller and approaches zero, we can determine that the instantaneous velocity is $16(2t) = 32t$. This means you know the speed of my pizza at any instant in time. If the pizza hits my hands after 1 second of falling from the ceiling, the pizza is traveling at 32 feet per second. (And you thought it was *easy* to catch a falling pie?)

I hope I have made it clear that $f(t) = 16t^2$ is the formula that describes the location of the pizza in feet. I hope I have made it equally clear that from $f(t) = 16t^2$ we can obtain another formula for the speed $f'(t) = 32t$. This new function f' is called the *derivative* of the first function f. We will talk more about derivatives in future lessons.

Calculus reminds me of a pizza chef standing on his or her two legs and tossing a pizza into the air. Calculus also has two legs. One leg is called the derivative. The other is called the integral. By the time you finish this book, both your legs will be strong.

Incidentally, pizza tossing and baking are merely metaphors for all the applications of calculus. Today calculus has invaded every field of scientific endeavor and plays invaluable roles in biology, physics, chemistry, economics, sociology, engineering—and in any field where some quantity, like speed or temperature, changes. Calculus can be used to help explain the structure of a rainbow, teach us how to make more money in the stock market, guide a spacecraft, make weather forecasts, predict population growth, design buildings, quantify happiness, and analyze the spread of AIDS.

Calculus has caused a revolution. It has changed the way we look at the world.

Derivative Definitions and Rules

2

August 1851: The night was chill [at the museum ball], and I dropped too suddenly from differential calculus into ladies' society, and I could not give myself freely to the change. After an hour's attempt to do so, I returned, cursing the mode of life I was pursuing. Next morning I had already shaken hands, however, with differential calculus and forgot the ladies.

—Thomas Archer Hirst, quoted in J. H. Gardner and R. J. Wilson, *American Mathematical Monthly*

Sometimes I write calculus formulas on napkins. The people who come to my restaurant are impressed by the way the formulas look. Perhaps you, too, can impress your friends and families.

We've already reviewed how to calculate the speed function from the distance function for falling pizzas. In other words, I gave you a formula that described where the pizza will be as a function of time, and together we computed the speed of the pizza at any location during its dramatic flight.

In general, we can determine a speed function from a distance function as follows. Consider a function $f(t)$. Let's write the derivative $f'(t)$ on a napkin. The derivative is a function that depends on the following ratio:

$$\frac{f(t+\Delta t)-f(t)}{\Delta t}.$$

As I explained, we want to determine the limiting value of this fraction as the time interval Δt gets so small that it vanishes. We can rewrite this formula to indicate our interest in this "vanishing" of time:

$$f'(t)= \lim_{\Delta t\to 0} \frac{f(t+\Delta t)-f(t)}{\Delta t}.$$

This is the theoretical definition of the derivative. The *lim* means that we want the value of this fraction in the limit as the time interval gets close to zero. If f represents the distance traveled, f' represents the instantaneous speed at a particular time. If f represents temperature, f' represents the instantaneous temperature change at a particular time. If f is any function, f' gives its rate of change. For example, if f gives the number of pizzas eaten in my shop through time, f' gives the rate of pizza eating, which might be one pizza per minute, especially on one of those hot Saturday nights.

The field of calculus is full of rules to make life easier. If you can remember just a few of these rules, you'll be a master and able to find the derivative of many functions. Some of my favorite rules are shown in the table below.

SOME SIMPLE DERIVATIVE RULES

Rule Name	Formula	Example
Power rule	$f(t) = t^n$ $f'(t) = nt^{n-1}$ (n is a real number)	$f(t) = t^4; f'(t) = 4t^3$
Multiplier rule	$(af)' = af'$ (a is a constant multiplier)	$(a3t^2)' = a(3t^2)'$
Sum rule	$(f + g)' = f' + g'$	$(3t^2 + 2t^4)' = (3t^2)' + (2t^4)'$

A lot can be accomplished with these rules. Remember all the hassle we went through in Chapter 1 when we analyzed the speed of the falling pizza? We had the formula for the distance traveled as a function of time, $f(t) = 3t^2$. We wanted the derivative of $f(t)$. That required quite a few steps.

However, now it is possible to easily find the derivative by using the multiplier rule followed by the power rule. Using the multiplier rule, we find that $(3t^2)' = 3(t^2)'$. And, using the power rule, this becomes $3(2t)$ or $6t$. Easy as pie! Note that when you use the power rule, t is just t. It can't be some complicated expression involving t. Notice also that the power rule tells us that the derivative of a constant is zero. You don't see why? Recall that any number raised to a 0 power is 1. So, for example, the derivative of a constant like 1,492 can be thought of as $(1492t^0)'$ or $(1492) \times 0 \times t^{-1}$, which is zero for $t \neq 0$. We expect the derivative of a constant to be zero because a constant has a zero rate of change.

There is a shuffling sound behind my restaurant counter. It is my assistant, Rosario. He has agile hands and is particularly good at kneading dough.

"Sir," he says, "*Ciao, che piacere vederti.* Can we have more examples of derivatives?"

"Of course. Find the derivative of $666t^4 - 2t^{100}$."

He nods. "This should be sufficiently impressive for the customers."

"Let's start," I say. "Using the sums rule, we search for $(666t^4)' + (-2t^{100})'$. Using the multiplier rule, this becomes $666(t^4)' + (-2)(t^{100})'$. Using the power rule, we obtain $(4 \times 666t^3) + (-100 \times 2t^{99})$, or $2664t^3 - 200t^{99}$. From a practical viewpoint, if $666t^4 - 2t^{100}$ describes the distance traveled by the pizza in millimeters as a function of seconds as the pizza glides through space (what a strange pizza!), then its speed at any time is $2664t^3 - 200t^{99}$ millimeters per second."

Rosario brings me a large pepperoni pie. "Sir, would you like a slice?"

"Certainly. *Mamma mia*, it is delicious." I pause to catch my breath. "Rosario, stick your finger in the center of the pie."

He quickly touches the pie and withdraws. "Sir, very hot!"

"I've been telling you about changes in distance with respect to time, t. Now, let's consider the change in temperature T of the pie as a function of the distance d from the hot center. Assume T is measured in degrees Fahrenheit and d is in inches. I'll make up an equation for this." I write on a napkin

$$T(d) = 150 - d^2.$$

"Rosario, what is the rate of change in the pizza's temperature 3 inches from the pizza's center?"

"Well," Rosario says, "first I find the derivative, which is $T'(d) = -2d$. When you probe the pizza 3 inches from its center, the rate of temperature change is -6 degrees per inch."

"Correct. The pizza's temperature decreases quickly as you examine it at increasing distances from the center. For example, at 4 inches from the center, the rate of temperature change is –8 degrees per inch. The negative sign means that the pizza is cooling."

"Sir, your formula suggests that the pizza is only 6 degrees Fahrenheit if you examine the pizza 12 inches from the center. That's below freezing!"

"Yes, Rosario. You can see that my initial guess at a temperature formula for the pizza was pretty crude. Your assignment is try to think of a more realistic formula. However, I think we could write a great science fiction story in which pizzas, for some unknown reasons, cool according to strange, unrealistic formulas."

"Cosmic," Rosario says.

"These pizzas would be very difficult to eat. But we are digressing—this is not the main point of our discussion."

■ ■ ■

For the next half hour, I remind Rosario that calculus deals with changes that don't have to be with respect to *time*. For example, we saw with the hot pizza how calculus could be used to study temperature change with respect to *position*. Calculus could be used to study pressure, intensity, or even levels of happiness with respect to locations. Therefore, we don't have to use our Δt notation, which focuses our attention exclusively on changes in time. Instead we often express quantities more generally like $y = f(x)$. Instead of writing $f(t + \Delta t) - f(t)$, as we did earlier today in

$$f'(t) = \lim_{\Delta t \to 0} \frac{f(t + \Delta t) - f(t)}{\Delta t},$$

we might rewrite this more generally as

$$f'(x) = \lim_{h \to 0} \frac{f(x + h) - f(x)}{h},$$

where h and x can be any quantities, for example, time, temperature, brightness, pressure, distance, electrical charge, state of arousal, level of happiness, number of people, darkness of hair color, toenail length, pizza consumption, or anything else that changes.

In fact, we can also write the derivative formula in a simpler fashion. The numerator becomes Δf. Then the derivative is

$$f'(x) = \frac{d}{dx} f(x) = \lim_{\Delta x \to 0} \frac{\Delta f}{\Delta x}.$$

Or, we can write this even more compactly using the notation of my idol, Gottfried Wilhelm von Leibniz, who invented calculus with Isaac Newton in the late 1600s:

$$f'(x) = \frac{df}{dx}.$$

In our hot pizza example, df is the tiny or infinitesimal change in temperature, and dx is the infinitesimal change in the distance from the center. In other words, df/dx is the instantaneous change in the temperature at any point in the pizza. Yet another way of writing the derivative of $y = f(x)$ is $y' = dy/dx$.

That's enough for today. *Ciao. A presto.* Come again tomorrow. I want to introduce you to my friends, Big Tony and Fiona.

EXERCISES....................

1. Rosario is a rocket hobbyist. He wants to earn a spot in the *Guinness Book of World Records* by being the first person to launch a meatball miles into the air (Figure 5). Rosario's crazy model rocket has a complicated set of rocket boosters. He believes that the distance of the meatball in miles from Earth can be represented as a function of time in seconds: $f(t) = 3t^2 + 4t^3$. How fast is the meatball moving after it has traveled 100 seconds? Do you think Rosario can actually get his meatball to travel this long and this fast?

2. Luigi has placed a pizza on a heating plate. The change in temperature T of the pie (in degrees Fahrenheit) varies as a function of the distance d (in inches) from the hot center. This temperature function is $T(d) = 160 - \frac{1}{2}d^2$. What temperature is the pizza half an inch from the center of the pie? What is the temperature change at this point?

FIGURE 5. Meatball in outer space.

3. Rosario has tried to derive a function to estimate the number of people in the pizza restaurant as a function of time. The time is measured in minutes after 7 o'clock on Saturday night, and the function is

$$f(t) = \frac{1}{10,000,000}(5t^2 + 3)^2.$$

(a) About how many people are in the restaurant at 8 o'clock at night? (b) At what rate are the people entering the restaurant at 8 o'clock?

4. Compute the derivative $y' = dy/dx$ for $y = 3x^2 - 3x + 2$.

5. Compute the derivative $y' = dy/dx$ for $y = 4x^3 - x/2 + 20$.

6. Compute the derivative $y' = dy/dx$ for $y = 1,000,000$.

7. Compute the derivative $y' = dy/dx$ for $y = \pi^2$.

8. Compute the derivative $y' = dy/dx$ for

$$y = \frac{1}{2}\pi^2 x - \frac{\pi^3}{3}.$$

9. Compute the derivative $y' = dy/dx$ for

$$y = \frac{\pi^3 x}{3} - \frac{ax^3}{2},$$

where $a = 4$.

10. Compute the derivative $y' = dy/dx$ for

$$y = \frac{\pi^3 x^2}{3} - \frac{ax^4}{2},$$

where $a = 4$.

3

Derivatives and Slopes

Throughout the 1960s and 1970s, devoted Beckett readers greeted each successively shorter volume from the master with a mixture of awe and apprehensiveness; it was like watching a great mathematician wielding an infinitesimal calculus, his equations approaching nearer and still nearer to the null point.

—John Banville, *New York Review of Books*

At dinnertime, my restaurant is packed with local factory workers talking shop over plates of steaming linguine and calamari, or Luigi's specialty—shrimp parmesan atop broccoli, mushrooms, and smoked green olives. A few of the tables are occupied by artists who exhibit their works in the gallery next door. These patrons really enjoy their meals after taking off their fringed vests and rolling up the sleeves on their Nehru jackets. An old-fashioned wood-and-wickerwork fan spins overhead. It creates a pleasant change from the arctic air-conditioning in many of the nearby stores and hotels.

Today I am drawing a curve, $y = f(x)$, on a wall of my restaurant (Figure 6). I turn to my customers. "How steep is the curve at the point I have marked? In other words, what is the slope of the curve at this point?"

Big Tony, one of my regular customers, says, "Luigi, I'm not sure. I
know how to calculate the slope of a
straight line. A line's slope is constant.
I just calculate the change in y for a
certain change in x. The more y
changes for a certain change in x, the
steeper the slope." Tony writes on the
wall

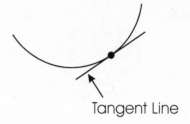

Tangent Line

FIGURE 6. Parabola with one
tangent line drawn.

slope of a straight line

$$= \frac{y_1 - y_0}{x_1 - x_0} = \frac{\Delta y}{\Delta x}.$$

I nod. "Big Tony, that's correct, and you also know that the equation of
a straight line is $y = mx + b$, where m is the slope and b is the y-intercept.
However, a curve such as the one I have drawn has a slope that's always
changing. At some places it is steeper than others. In fact, what we want
to determine for a curve is the *limiting ratio* of a change in y given a
tiny change in x. This ratio is just what the derivative is." I write on the
wall

$$\frac{\Delta y}{\Delta x} \Rightarrow \lim_{\Delta x \to 0} \frac{f(x + \Delta x) - f(x)}{\Delta x} = \frac{dy}{dx}.$$

For a moment, I look at Big Tony and I think of the past. I have known
Big Tony since we were teenagers. We lived on the same street. Although
we had our differences, we nevertheless were friends, probably because
our senses of humor were similar and we both liked reading Italian cook-
books.

I signal for the waitress. In contrast to Big Tony, Fiona is slim with sharp
cheekbones and a scraggly mop of long, tangled maroon hair. Three silver
earrings with little alien heads protrude from her left ear. "All You Need Is
Love" is stenciled on her shirt beneath a peace sign. Glass necklaces and
wooden beads surround her neck like adoring snakes.

Fiona points to the parabola on the wall (Figure 6). "Let's assume that
this is a drawing for $y = f(x) = x^2$. The derivative is $f'(x) = 2x$. The steepness

at any point along the parabola keeps changing." She moves her chartreuse fingernails along the parabola. "At the very bottom of the parabola, it isn't steep at all! In fact, the derivative is zero at this point. From a geometrical standpoint, the derivative gives us the slope of the tangent line to $f(x) = x^2$ at a point on the curve."

"Ooh!" I say. "Fiona, you've been studying."

She smiles.

Big Tony puts his hands in his pockets. "Luigi, what exactly is a tangent line?"

"Look at the figure," I say. "Notice that I drew an example—a line that goes through a given point and travels in the same direction as the curve at that point. The tangent line just kisses the curve at the point. Think of the tangent line as a seesaw plank that pivots according to the curve's steepness."

Big Tony nods. "Right. At $x = 0$, the parabola is flat and the tangent line is horizontal. At $x = 1$, the derivative is 2, so the slope is 2. At $x = 2$, the derivative is 4, implying that the curve is getting ever steeper."

"Big Tony," I say, "let's try to find where the slope is zero for a more complicated curve, $y = f(x) = x^2 - 4x - 6$. Taking the derivative of $f(x)$, we find $f'(x) = 2x - 4$. You can see that the slope is 0 when x is 2. Big Tony, can you find the equation for the line tangent to this graph at point $(1, -9)$?"

Tony gulps down his huge slice of pizza loaded with anchovies, onions, and gefilte fish. "We know that the slope at any point on the curve is $2x - 4$, so this means the slope of the tangent line at $x = 1$ is -2."

"Correct," I say.

Big Tony continues. "The equation of a line with slope m is $y = mx + b$. So, the line's equation is $y = -2x + b$. We can solve for b by inserting any point on the curve into the equation. Let's try $(1, -9)$, which gives us $-9 = -2 \times 1 + b$, so b is -7. Given our original equation, $f(x) = x^2 - 4x - 6$, the equation of the tangent line at $(1, -9)$ is $y = -2x - 7$."

Fiona claps her hands and gives Big Tony a thumb's up.

From the back of the kitchen I hear Rosario shouting, "We're out of anchovies and olives, and our customers won't like that one bit." Pandemonium ensues as pots and pans drop to the floor.

I turn to Fiona and Big Tony. "You can see that I am needed in the kitchen. Let's meet again tomorrow."

EXERCISES.........................

1. Big Tony wants to know if the graph of $y = x^3 - 1$ is steeper at $x = 2$ or at $x = -3$. (a) What is the slope of the tangent line at these two points? (b) At which point is the slope larger? (c) Write the equation for the line tangent at $x = 2$.

2. What is the equation of the line tangent to the curve $y = x^2 + 10$ at $x = 2$? Is the curve steeper at $x = -2$ or $x = 1$? Why?

3. Given the curve defined by $y = x^2 - x + 6$, where is the curve least steep?

4. If you were to roll a meatball down the curve defined by $y = -x^2 + x + 10$, at which point would the meatball eventually settle? (Hint: Where is the slope equal to zero?)

5. For the curve with the equation $y = \pi x^5 - x$, find the slope of the tangent at $x = 0$.

6. For the curve with the equation $y = \pi x^6 - x^2$, find the slope of the tangent at $x = 0$.

7. For the curve with the equation

$$y = \frac{\pi \sqrt{x}}{2},$$

find the slope of the tangent at $x = 1$. (Hint: Recall that another way to write $\sqrt{x}$ is $x^{1/2}$.)

8. For the curve with the equation

$$y = \frac{\pi \sqrt{x}}{2} + 3x^2,$$

find the slope of the tangent at $x = 1$.

9. For the curve with the equation $y = (x + 2)^2$, find the slope of the tangent at $x = 0$.

10. For the curve with the equation $y = (x + 3)^2$, find the slope of the tangent at $x = 0$.

4

Rules for Products and Quotients

Do you mean you talk calculus with women you're NOT attracted to?

—Dick Solomon, *Third Rock from the Sun*

nything new with you, Fiona?"

"Yes, Luigi," she says. "I'm taking a course Tuesday nights—conversational Italian. I figure it's about time I learned."

"Good for you!" I say.

Fiona is a bright woman—it's one of the reasons I find her fascinating. And she is so easygoing. She's certainly different from her father. I met her parents once when they invited me over for dinner. Everything started precisely on time. Her father ruled with an iron will. Reminded me of a sergeant I once knew during a brief stint in the military.

"Fiona, how would you find the derivative of this impressive formula?" I hand Fiona a card containing the following terms:

$$(3x^5 + 2x^3 - 7x)(x^4 + x^3 + 3x^2 + 1).$$

Fiona starts to shake as she tries to multiply the two big functions.

"Fiona, stop right there! You don't have to do the massive multiplication."

"Thanks," she says uncertainly as she rubs a slight scar near her left eye. The scar is a squiggly mark that looks like an infinity symbol. But then her body seems to relax and her playful smile is back.

"Let me give you two new derivative rules, in addition to the rules I gave Rosario in Chapter 2."

"Luigi, why do you say 'Chapter 2'? You're acting like we're in some kind of book."

I wave my hand. "Never mind that. Take a look at the new rules scribbled on the ceiling of the restaurant."

Fiona looks up and sees the product rule, quotient rule, and their formulas.

Rule Name	Formula
Product rule	$(fg)' = f'g + fg'$
Quotient rule	$\left(\dfrac{f}{g}\right)' = \dfrac{f'g - fg'}{g^2}$

"Fiona, if you remember anything from today's lesson, remember that the derivative of the product of two functions $(fg)'$ is *not* the product of the derivatives $f'g'$. Now, let's use the product rule to figure out the derivative of our massive formula." I write on the wall

$$(3x^5 + 2x^3 - 7x)(x^4 + x^3 + 3x^2 + 1)$$

Let $f = (3x^5 + 2x^3 - 7x)$ and $g = (x^4 + x^3 + 3x^2 + 1)$

Let $f' = 15x^4 + 6x^2 - 7$ and $g' = 4x^3 + 3x^2 + 6x$.

Therefore, using the product rule,

$$y' = (fg)' = (15x^4 + 6x^2 - 7)(x^4 + x^3 + 3x^2 + 1)$$
$$+ (3x^5 + 2x^3 - 7x)(4x^3 + 3x^2 + 6x).$$

I snap my fingers. "Fiona, you can see that we did not have to multiply the two big functions first to get the derivative. That makes our lives much easier."

Fiona runs her fingers through her hair. "Luigi, yes, that's a very helpful trick."

"Just by looking at our result, I can tell you that at $x = 0$, the derivative is $(-7)1 + 0$, or -7. Who knows what this curve $(3x^5 + 2x^3 - 7x) \times (x^4 + x^3 + 3x^2 + 1)$ looks like, but we can find the derivative with the product rule without making a plot."

Big Tony is sitting at a nearby table. He watches Fiona and me as he waits for his entree.

"Luigi," Fiona says, "let me try to apply the quotient rule to your two big formulas, and in this case I'll compute the derivative for $(3x^5 + 2x^3 - 7x) \div (x^4 + x^3 + 3x^2 + 1)$."

Big Tony heaves himself up from the table, tosses aside some garlic bread, and starts scribbling on the tablecloth

$$\left[\frac{(3x^5 + 2x^3 - 7x)}{(x^4 + x^3 + 3x^2 + 1)} \right]'$$

$$= \frac{(15x^4 + 6x^2 - 7)(x^4 + x^3 + 3x^2 + 1) - (3x^5 + 2x^3 - 7x)(4x^3 + 3x^2 + 6x)}{(x^4 + x^3 + 3x^2 + 1)^2}.$$

Fiona helps Tony by copying the final result on the wall using her red lipstick.

A few of the restaurant customers gaze at the magnificent equation and begin to applaud. The cigar dangling from Big Tony's lip falls to the floor, and Fiona takes a bow.

My heartbeat increases in frequency and amplitude as a few of the patrons pick me up on their shoulders shouting, "We want more calculus! We want more calculus!"

Just another typical evening at Luigi's Pizza and Calculus Restaurant.

 **EXERCISES**....................

1. Big Tony wants to know the derivative of

$$\frac{x}{x^3 - 1}.$$

Can you help him by using the quotient rule?

2. Use the quotient rule to find the derivative of

$$\frac{x^2 + 2x + 2}{x^2 + 1}.$$

3. Find the derivative of $y = (x^2 + 1)(x^2 + 2)$ using the product rule. What is the slope of the curve at $x = 0$?

4. Find the derivative of $y = (\frac{1}{2}x^3 + 1)(x^4 + x^2 - 100)$ using the product rule. What is the slope of the curve at $x = 0$?

5. Use the product rule to find the derivative of $y = (\sqrt{x} + 1)(x^2 + 2x + 2)$.

6. Use the product rule to find the derivative of $y = (2\sqrt{x} + 1)(x^2 + 2x - 2)$.

7. Use the quotient rule to find the derivative of

$$y = \frac{x}{\pi}.$$

8. Use the quotient rule to find the derivative of

$$y = \frac{x^2}{\pi}.$$

9. Use the product rule to find the derivative of

$$\frac{1}{x}\sqrt{x}.$$

(Hint: Recall that $1/x$ can be rewritten as x^{-1}.)

10. Use the product rule to find the derivative of

$$\frac{\pi}{2x}\sqrt{x}.$$

5

Chain Rule and Implicit Differentiation

The body of mathematics to which the calculus gives rise embodies a certain swashbuckling style of thinking, at once bold and dramatic, given over to large intellectual gestures and indifferent, in large measure, to any very detailed description of the world. It is a style that has shaped the physical but not the biological sciences, and its success in Newtonian mechanics, general relativity and quantum mechanics is among the miracles of mankind.

—David Berlinski, *A Tour of the Calculus*

I stop chewing for a second and then fork a big fat shrimp. "Tastes good," I say to Big Tony, Fiona, and Rosario. I pause for a few seconds to enjoy the zesty dish. "Today we are interested in the *chain rule* that allows us to differentiate more complicated functions, including functions of functions. Consider, for example, the following equation." I write on a napkin

$$y = (x^3 + x + 1)^{25}.$$

"The derivative would be a real monster to calculate if we had to multiply $(x^3 + x + 1)$ twenty-five times before finding the derivative."

"*Mamma mia!*" Rosario shouts. "That's a lot of multiplying."

I nod. "Luckily for us, the chain rule says that if $y = f(u)$ and $u = g(x)$, then . . ."

$$\frac{dy}{dx} = \frac{dy}{du} \times \frac{du}{dx}.$$

Fiona picks up the napkin and runs her fingers through its soft folds. "Isn't that a beautiful formula?"

"What the heck does this mean?" Big Tony says.

"Let's look at $y = (x^3 + x + 1)^{25}$," I say. "We want to find the derivative. Let's make u equal to $(x^3 + x + 1)$":

$$u = (x^3 + x + 1).$$

I look at Big Tony. "This simplifies our equation so that it can be recast as . . ."

$$y = u^{25}.$$

"Big Tony, to find the derivative we first compute the derivative of y with respect to u, a process that may be written as dy/du."

"The derivative of y is $25u^{24}$," Fiona calls out.

I nod. "This is dy/du in the chain rule. We also take the derivative of u with respect to x, which is du/dx. This yields $3x^2 + 1$. Using the chain rule, we obtain the following":

$$\frac{dy}{dx} = \frac{dy}{du} \times \frac{du}{dx} = 25u^{24} \times (3x^2 + 1).$$

"Finally, we substitute the true value of u, which is $(x^3 + x + 1)$, so that . . ."

$$\frac{dy}{dx} = 25(x^3 + x + 1)^{24} \times (3x^2 + 1).$$

"Excellent," says Fiona. "We now have the derivative."

And I explain, "Yes, and although the derivative looks complicated, the work we had to do to find the derivative was pretty easy because we had the chain rule to work with.

"Notice that our initial example, $y = (x^3 + x + 1)^{25}$, might be considered a function of a function. One function is $(x^3 + x + 1)$, and the other function raises this to the 25th power. So, another way of thinking about the chain rule is to think of it applied to a function f of a function g. This suggests another way of writing the chain rule":

$$\frac{d}{dx} f\big(g(x)\big) = f'\big[g(x)\big]g'(x).$$

"In other words, the derivative of a function of a function (also known as a composite function) is the derivative of the outside function f evaluated at the inside function g, times the derivative of the inside function g. In our example, the raising to the 25th power was the outside function."

"Good job, sir," Rosario says.

For a moment I am distracted as my gaze wanders to various sculptures in my restaurant window. These are all gifts from Big Tony. Quite impressive. Toward my right is a marble head of the god Jupiter, a Roman sculpture from the first century A.D. Behind the head is an Etruscan vase dating to circa 490 B.C. The vase depicts Heracles fighting the Nemean Lion. After damaging his weapons on the lion's impervious hide, Heracles must choke the beast to death.

Fiona follows my gaze. "Big Tony always has expensive and unusual tastes."

Big Tony grins as he tosses a meatball to Fiona.

"Hey, stop that. Are you nuts?"

"What?" Big Tony says. "You're not a vegetarian?"

I clap my hands together. "Stop playing. Next, I want to tell you about the *extended power rule*." As I write these words on the wall of the restaurant, a shiver goes up my spine. There's nothing like good mathematics. I hope Fiona can appreciate that.

I turn to Fiona. "Remember the power rule we learned? It looked like this." I write on the wall

$$(x^r)' = rx^{r-1}.$$

"Yes," she says. "For example $(x^3)' = 3x^2$."

"Yes. Using the chain rule, I can now present the extended power rule":

$$(u^r)' = ru^{r-1}u'.$$

I look at Big Tony, then at Fiona. "Here r is a real number constant. It can't be a variable. As an example, let's find the derivative of the cube root of $2x + 1$, which may be written as . . ."

$$\left(\sqrt[3]{2x+1}\right)'.$$

Big Tony grabs the marker from my hand. "Which can also be written as . . ." He writes on the wall

$$\left[(2x+1)^{\frac{1}{3}}\right]'.$$

I grab the marker back from Tony's huge hand. "Let $u = 2x + 1$ and $r = 1/3$. Using the extended power rule, we get . . ."

$$u = 2x + 1 \qquad r = 1/3$$

$$u' = 2 \qquad r - 1 = -2/3$$

$$(u^r)' = ru^{r-1}u' = \frac{1}{3}(2x+1)^{-\frac{2}{3}} \times 2.$$

"Recalling some rules from algebra, such as $x^{p/q} = \sqrt[q]{x^p}$ and $x^{-r} = \dfrac{1}{x^r}$, we can rewrite this as . . ."

$$\frac{2}{3}\frac{1}{(2x+1)^{\frac{2}{3}}} = \frac{2}{3\sqrt[3]{(2x+1)^2}}.$$

"Let's work through another problem."

"Sir," Rosario says, "I love learning through examples."

"Consider this one, which makes use of the extended power rule. Let's find the derivative for this:"

$$\left(\sqrt{\pi x^3 - 4x + \pi}\right)'.$$

"To start, note that this can be rewritten as . . ."

$$\left[(\pi x^3 - 4x + \pi)^{\frac{1}{2}}\right]'.$$

"Let $u = \pi x^3 - 4x + \pi$. Using the chain rule and extended power rule, we find . . ."

$$\frac{1}{2}(\pi x^3 - 4x + \pi)^{-\frac{1}{2}}(\pi x^3 - 4x + \pi)'$$

$$\frac{1}{2}(\pi x^3 - 4x + \pi)^{-\frac{1}{2}}(3\pi x^2 - 4)$$

$$\frac{1}{2}\frac{1}{\sqrt{(\pi x^3 - 4x + \pi)}}(3\pi x^2 - 4)$$

$$\frac{(3\pi x^2 - 4)}{2\sqrt{(\pi x^3 - 4x + \pi)}}.$$

From within the subterranean depths of his linguine, Big Tony uncovers a meatball in the shape of the Statue of Liberty. It must be a joke being played on him by the new chefs, Britney and Christina. "I must remember to compliment them. Just how do they do it?" asks Big Tony.

"Let's continue," I say. "We can also use the chain rule for trigonometric functions like sine and cosine."

Fiona raises her hand. "Luigi, we haven't discussed derivatives for those yet."

"Correct. Let me teach you about how to treat trigonometric functions. You should all recall what sine waves and cosine waves look like." I show them my sine wave and cosine wave tattoos (Figure 7).

"Now, let me tell you about the derivative rules for sines and cosines."

Trigonometric Derivatives

$$(\sin x)' = \cos x$$
$$(\cos x)' = -\sin x$$

"And, similarly, we have the sine and the cosine chain rules."

Trigonometric Chain Rules

$$(\sin u)' = (\cos u)u'$$
$$(\cos u)' = -(\sin u)u'$$

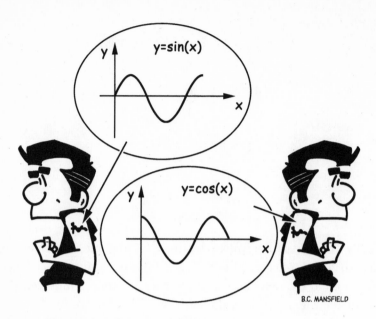

FIGURE 7. Sine and cosine wave tatoos.

"One way to visualize these relations is to notice that sine and cosine waves are out of phase. For example, when the sine function is at a maximum or a minimum and the slope is zero at these points, the value of cosine is indeed zero."

"Luigi," Fiona says, "Can we try a few examples?"

"You bet. Find the derivative of $100(\sin x)$."

She nods. "According to the first trigonometric rule you gave us, the derivative is $100(\cos x)$."

"Correct. Find the derivative of $\cos(10x^2)$."

"Luigi, using the chain rule, we let $u = 10x^2$. So, we get $-(\sin 10x^2)(10x^2)'$ or $-\sin(10x^2)(20x)$."

"Excellent."

I hand Big Tony a card with two words on it:

IMPLICIT DIFFERENTIATION

"Sounds cool," Big Tony says.

"Very. Ready to learn more?"

"Bring it on, Luigi."

"Sometimes we will have an equation for which we want the derivative dy/dx, but we will find it inconvenient to actually cast the equation so that it looks like $y = f(x)$."

"Sir," Rosario says, "give us an example."

"Consider $y^3 - xy + 7x = 15$. Even though this is not explicitly written in our usual format of y as a function of x, we can still differentiate the entire equation with respect to x. It's a lot of fun."

"Great," says Fiona.

"Here's an example. Our mission is to find the derivative dy/dx if y is defined implicitly (or indirectly) as a function of x by the following equation." I write on Fiona's right arm, directly beneath her purple monarch butterfly tattoo,

$$y^3 - xy + 7x - 15 = 0.$$

"Fiona, let's pay attention to the y^3 part of the equation. We represent its derivative by $3y^2(dy/dx)$."

Fiona closes her eyes for a second. "Luigi, explain."

"I used the chain rule to find the derivative of y^3. Let's assume we don't know the derivative of y—after all, we have not put this equation in the form where y is on one side of the equals sign—so wherever we encounter the derivative of y, we just write dy/dx. Remember, by the chain rule, the derivative of u^3 is $(u^3)' = 3u^2u'$. Applying this to y itself, we get $(y^3)' = 3y^2y'$."

Fiona moves my felt-tip pen from her arm. "What's the derivative of the xy part of the equation?"

I write on the wall. "We can use the product rule, which says $(fg)' = f'g + fg'$. Let f be x and g be y. So, we get $1 \times (y) + x \times dy/dx$. The last term, $7x$, has no y, so its derivative is simply 7. Putting this all together we get the following expression":

$$3y^2\frac{dy}{dx} - 1y - x\frac{dy}{dx} + 7 = 0.$$

"Next we just solve this equation for dy/dx":

$$3y^2 \frac{dy}{dx} - x \frac{dy}{dx} = y - 7$$

$$\frac{dy}{dx}(3y^2 - x) = y - 7$$

$$\frac{dy}{dx} = \frac{y - 7}{3y^2 - x}.$$

Fiona turns to Rosario. "Isn't that wonderful, Rosario? We calculated the derivative of y without ever explicitly solving for y in our complicated equation."

Big Tony looks at me. "What if we wanted to find the slope of the curve at a particular point?"

"Easy. The trick is that we have to insert both x and y values—not just an x value —to find the slope at a point. For example, at the point ($x = 2$, $y = 0$), we have a slope of 7/2."

I turn to my adoring colleagues. "In the coming days, I'm going to show you how you can apply all of this basic knowledge to solve some real problems with pizzas and Italian cooking, including my favorite problem of the growing meatball."

"Ooh," Fiona says. "That sounds very weird, but delicious."

"But now it is time to rest. See you all tomorrow."

∎ ∎ ∎

I relax in one of the back rooms of the restaurant. Three species of olive trees stand against the tall glass windows. The magnificent plants vary from small to giant, black to brilliant green, gold with green stripes, and green with cream stripes. Some are straight like telephone poles with large leaves; others have bulging Buddha belly profiles with clouds of fine leaves.

Next time I should really ask Fiona back here. She would like the peaceful setting.

I smile at the Greco-Roman statue of Bacchus, the god of wine, whom the Greeks called Dionysus. He was a noble god who taught humans how to cultivate grapes and make wine. The Bacchanalia, or the festival of Bacchus, was celebrated every third year, but it became so wild and uncontrolled that in 186 B.C. the Roman Senate forbade future celebrations.

For a moment I just sit on an oak bench in the room, enjoying the sharp, welcome scent of olives coming through the window. A gentle breeze is blowing outside, and all around me, I hear the murmur of the olive leaves and stalks. It reminds me of the whispers of the ancient gods.

EXERCISES.........................

1. Use the chain rule to compute the derivative dy/dx of

$$y = \left(\frac{x^4}{4} + 4 \right)^4.$$

What is the slope at $x = 1$?

2. Use the chain rule to compute the derivative dy/dx of

$$y = \left(\frac{x^3}{\pi} + \pi \right)^3 - \left(\frac{x^2}{\pi} + \pi \right)^2$$

(Hint: Just treat π as a constant.) What is the slope at $x = 1$?

3. Use implicit differentiation to compute the derivative dy/dx for $x^3y - (y + x)^4 + 1 = 0$. What is the slope at $(0, 1)$?

4. Use implicit differentiation to compute the derivative dy/dx for $ax + by = 0$, where a and b are constants. What is the slope at $(-b, a)$?

5. Use the extended power rule to compute

$$\left(\frac{\pi}{x^2 + 1} \right)'.$$

What is the slope at $x = 1$?

6. Using the extended power law and any other tricks you know, find the derivative of

$$\frac{\pi}{\sqrt{x^2+\pi}} + \sqrt{\sqrt{x}}.$$

7. Use the trigonometric and chain rules to find the derivative dy/dx of $y = \pi\cos(3x^3)$. What is the rate of change of this function at $x = 0$?

8. Use the trigonometric and chain rules to find the derivative dy/dx of $y = 4\sin(x^3 + 2x^2 - 1)$.

9. Use the chain rule to find the derivative of

$$y = \sqrt{1 + \sqrt{1 + \sqrt{1 + x}}}.$$

Isn't this a beautiful equation?

10. Use the chain rule to find the derivative of

$$y = \sqrt{2 + \sqrt{1 + x}}.$$

Maxima and Minima

> Mathematics is the instrument by which the engineer tunnels our mountains, bridges, our rivers, constructs our aqueducts, erects our factories, and makes them musical by the busy hum of spindles. Take away the results of the reasoning of mathematics, and there would go with it nearly all the material achievements which give convenience and glory to modern civilization.
>
> —Edward Brooks, *Mental Science and Culture*

I walk across the patio that leads to my restaurant's front door. The ground contains moss-covered stones, and to the left is a wishing well where people are invited to toss a coin if they have any culinary wishes left unsatisfied. There are no coins in the well. Of course not. My pizza is the best in the world.

Fiona, Rosario, and Big Tony are waiting for me inside. Fiona calls to me. "Luigi, we are ready for the next calculus lesson. Come join us and eat while you teach." She motions to some antipasto on the table.

I stare at a Luigi special—a platter of Milan prosciutto, spicy capicola, aged provolone, calamata olives, roasted peppers and mushrooms, marinated artichoke hearts, grilled eggplant, roasted red onions, and fresh mozzarella, all served with garlic bread drizzled with my homemade balsamic vinaigrette. More food will arrive soon.

"Let's begin," I say as I pop a pepper into my mouth. "Today I want to talk about the shape of curves. Imagine a curve defined by $y = f(x)$. If the derivative $f'(x)$ is greater than zero at a point on the curve, then the function is increasing at the point x and the slope of the tangent line is positive. If the derivative $f'(x)$ is less than zero at a point on a curve, then the function is decreasing at point x and the slope of the tangent line is negative." I draw on the wall (Figure 8).

I turn to the customers in the restaurant and speak loudly. "I am happy to announce that for the next hour the function that describes the prices you will pay for pizzas has a derivative that is negative."

The customers are quiet and stare at me. Rosario coughs nervously.

I feel a shiver go up my arm as I look into Rosario's shiny face. I feel a chill, an ambiguity, a creeping despair. It must be the silence of the customers. They don't move. Big Tony's smile is relentless and practiced. For a moment, the grilled eggplant on the antipasto seems to wriggle with life. But when I shake my head, the vegetables are still. However, Big Tony's smile remains. Cruel. Nightmarish. I feel like I am a gladiator forced into combat at a Roman coliseum and all the torch lights are out.

I turn to my customers. "C'mon people," I say. "A negative derivative means the slope of the curve is decreasing. It means pizzas will become less and less expensive for the next hour."

Everyone suddenly breaks out in applause. Big Tony comes up to me and slaps me on the back. "Good one," he says.

Rosario whispers, "What if the derivative of a function is exactly zero at a particular point on the curve?"

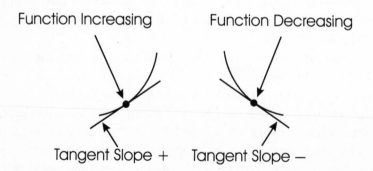

Function Increasing   Function Decreasing

Tangent Slope +   Tangent Slope −

FIGURE 8. A function that increases and a function that decreases.

"It means that the slope is zero. The curve isn't increasing or decreasing. The tangent is horizontal. We call the point on a curve that has a zero derivative a *critical point*. Think of a bottom point of a parabola at the origin (0, 0). The slope of the curve is decreasing until it bottoms out at (0, 0) at which point the slope is zero. After that, for positive x values, the parabola has a slope that is increasing."

Fiona raises her hand. I notice several large blue-and-yellow triangles on her shirt. Her shirt's buttons are shaped like spirals, and her earrings—light-emitting diodes—are blinking. "How about a mountain analogy?" she asks. "Imagine I am at the base of a round-topped mountain, such as the geologically older and worn Green Mountains in Massachusetts. I begin my ascent, and the slope starts to slowly increase. Somewhere near the top of the mountain, the slope is still positive but starts to decrease. This is because as I come closer to the cap of the mountain, the land gradually becomes less steep, then it finally goes from a positive slope (indicating I'm going uphill) to a negative slope (indicating I'm going downhill) just as I come over the crest."

"Good. When $f'(x) = 0$, the curve may have a *local minimum*, a *local maximum*, an *inflection point*—or the curve might be flat such as would be the case for $f(x) = 10$." I draw examples of the first three cases (Figure 9). "Think of the third example, which is neither a local maximum nor a local minimum, as a resting point before continuing your ascent or descent along a mountain. It's called an inflection point. I'll talk more about inflection points later."

"Sir," Rosario says, "why do you use the word *local* when describing the minima and maxima?"

"Well, the curve can be very wiggly and can have many hills and valleys like the crests and troughs of ocean waves. There might be no *single*

Maximum   Minimum   Inflection Point

FIGURE 9. Points on a curve where the derivative is zero.

maximum and minimum, but there can be numerous local ones or relative ones." I draw a wiggly curve (Figure 10).

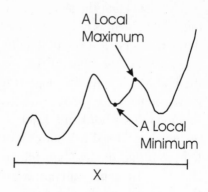

"Notice that in my figure the absolute maximum value for the range of x values shown is actually the rightmost point on the curve. And the absolute minimum value of the range of x values happens to be the leftmost point on the curve.

"An example will help make this clearer. Our task is to find all local maxima and local minima of the function $f(x) = x^2 + 2x - 3$."

FIGURE 10. An example of a local minimum and a local maximum for a range of x values.

Big Tony explains, a bread stick dangling from his mouth, "These maxima and minima can occur only at those locations of a curve where $f'(x) = 0$." He writes on the wall

$$f'(x) = 2x + 2.$$

Big Tony turns back and looks at Fiona. "The derivative $f'(x) = 0$ only when $0 = 2x + 2$ or when $x = -1$. This is a local minimum for the curve. Notice that for $x < -1$, the function is decreasing, and when $x > -1$, the function is increasing."

"Are you sure that point is at a minimum?" Fiona says.

"Yes," Big Tony responds. "Because at the point $x = -1$, $y = -4$, the graph of this function looks like a parabola with its minimum at $(-1, -4)$. Anyway, isn't this grand? We were able to tell that the curve has one local minimum and no local maximum."

I nod. "Very good work, Big Tony. Let me tell you about an easier way to determine for sure whether $x = -1$ corresponds to a minimum or maximum. We can tell simply by applying the *second derivative test*. If the second derivative is greater than zero at $f(x)$, then we have a local minimum at this point. If the second derivative is less than zero at $f(x)$, then we have a local maximum at this point."

Fiona takes a step closer to me. Her eyes are wide with astonishment.

"To find the second derivative," I say, "just take the derivative of the derivative. In our case, we have $f'(x) = 2x + 2$, so $f''(x) = 2$. The second derivative is the rate of change in the first derivative. In our example, the critical point is therefore a local minimum because the second derivative at this point is positive."

I draw a table on the wall in order to help Fiona, Rosario, and Big Tony remember all of the relationships between second derivatives and the nature of the critical point at x_c.

CRITICAL POINT AT X_c

Second Derivative	Type of Critical Point	Memory Aid
$f''(x_c) > 0$	Local minimum at x_c	⇑⇓
$f''(x_c) < 0$	Local maximum at x_c	⇓⇑

"What are the arrows?" asks Fiona.

"That's the simple way I remember the rule. When the derivative is big (⇑, i.e., greater than zero), we have a minimum (⇓). When the derivative is small (⇓, i.e., less than zero), we have a maximum (⇑)."

"Let's try another," I say. " Where are the local minima and maxima of the following?"

$$(1/3)x^3 - 9x - 2.$$

"Let me take a crack at this," Fiona says. "First we take the derivative, which is $x^2 - 9$. Now we find where the derivative is zero. Solving for x in $x^2 - 9 = 0$, we find that $x^2 = 9$, which is true for $x = 3$ and $x = -3$. This means we have critical points at these two values of x."

"Correct," I say.

Fiona nods and shuts off her earrings.

"Fiona, let's find out if these points are minima, maxima, or inflection points. To do so, we take the second derivative, which is $(x^2 - 9)'$ or $2x$. At the point $x = 3$, the second derivative is positive, so we have a minimum here. At the point $x = -3$, the second derivative is negative, so we have a maximum here. One of your exercises tonight, Rosario, is to draw the graph of this curve to see if you can see the maximum and minimum."

"What happens at points where $f''(x)$ is exactly zero?" Fiona asks.

"At these points, the slope is neither decreasing nor increasing. For example, this might occur at an *inflection point* in the curve—the resting point for a mountain climber. These are points where a curve changes from concave downward to concave upward, or vice versa." I point to the inflection point in one of my previous diagrams (Figure 9).

"As an extra exercise, experiment with the curve $y = x^3$ and examine the behavior of the slope at $x = 0$. You'll see that the slope is zero at this point, but this curve has neither a maximum nor a minimum. For this curve, $f'(x) = 3x^2$ and $f''(x) = 6x$. At $x = 0$, the second derivative changes from negative to positive, so this is a point of inflection."

Fiona nods. "Must inflection points have both the first and second derivatives as zero?"

"No. The first derivative can have any value at an inflection point. In fact, $\sin(x)$ has an inflection point at $x = 0$ where the slope is 1. It's also possible for the second derivative to be zero without there being an inflection point. For example, $y = x^4$ has $y''(0) = 0$, but this curve is concave upward everywhere. Thus, $y = x^4$ does not have an inflection point at $x = 0$ even though $f''(0) = 0$." I hand Fiona a card that explains what we know about inflection points.

Everything You Wanted to Know about
Inflection Points but Were Afraid to Ask

- If the second derivative exists at an inflection point, it must equal zero.

- The first and second derivatives need not exist at an inflection point, but the function must be defined at an inflection point.

- To find the inflection points of a function $f(x)$, find all points where $f''(x) = 0$ and all points where $f''(x)$ does not exist but $f(x)$ is defined. To check whether each of these candidates is an inflection point, determine whether the sign of $f''(x)$ changes at these points.

I watch as Big Tony chomps on an anchovy and garlic pizza. I point my finger at him. "I should warn you that certain functions, like $f(x) = x^{2/3}$, have a critical point where the derivative doesn't exist."

"Sounds cosmic," Big Tony says as he waves his aromatic pizza slice at me. A waitress brings us more pizza.

"Yes, for example, the first derivative might be infinite because you have to divide by zero. If you plot $f(x) = x^{2/3}$, you'll see it has a minimum at $x = 0$, even though the derivative isn't defined at $x = 0$. After all, $f'(x) = (2/3)x^{-1/3}$, which forces us to divide by zero":

$$f'(0) = \frac{2}{3x^{\frac{1}{3}}} = \infty.$$

"Sometimes it really pays to plot functions to get a feel for their behavior."

Rosario rushes off to the kitchen and returns with my favorite sauce. I start to add a spoonful of Mamma Mia's Special Sauce to my pizza slice.

"What's it made of?" Fiona asks.

"Thai red peppers," I say. "If the pizza doesn't make you sweat, you're not really eating!"

"Can I try some?" Fiona asks.

I roll my eyes. "You don't want to."

"I can take it," Fiona says. When I hesitate, she asks, "What are you, some kind of superhero?"

"Okay, I warned you. Just take a little."

Fiona takes a spoonful of pepper sauce and dumps it on her pizza. She takes a bite.

"Oh my!" she chokes and grabs for the water.

"I told you to be careful. It takes getting used to."

Big Tony laughs. "Fiona, if you think that's hot, I love Nuoc Tuong sauce. It contains soybean paste, garlic, fish sauce, chili, lemon juice, sugar, nutmeg, and ginger. You mix it all together in a tiny wood barrel, bury the barrel in the ground, and let the sauce ferment for six months before putting it on the food. Now that's a taste of heaven!"

I nod. "Ah, a man after my own heart." I pull a red Thai pepper from my dish, chew on it, and use my napkin to wipe the trace of sweat from my brow.

Fiona sighs. "What is it with men and spicy food? Is it some kind of macho contest?" She pauses and says, "Luigi, you have the most bizarre restaurant in town."

I smile as I watch Fiona apply a strange zebra-striped lipstick to her lips. I say, "Let me conclude today's lesson by talking about *concavity*. Imagine you are an ant walking on the inside of a bowl. In fact, don't imagine it. Here's the real thing."

I find an ant outside and drop it into a bowl. "The second derivative is positive everywhere the ant walks in the bowl (except for the point right at the center of the bowl)" (Figure 11).

Fiona looks at the ant in the bowl. The ant has trouble climbing the sides.

I continue. "The same would be true if the ant were walking inside of a parabola. Wherever the second derivative is positive, we say the surface or curve is *concave up*. Notice that for this bowl, one side is going down and another side is going up. So we can have a function that is either *increasing* or *decreasing* in slope that is, at the same time, always concave up—like the bowl. Similarly, if we flip the bowl over, or turn a parabola upside down, and place the ant on top, the second derivative is negative, and the curve or surface is called *concave down*."

Fiona nods. "An upside-down bowl also can have a first derivative that is positive or negative, which implies a function that is increasing or decreasing."

"Correct. Here is a table about concavity and slope to help you remember what we've been discussing." I pull out a card and hand it to Fiona.

B.C. MANSFIELD

FIGURE 11. An ant walking on a surface that is concave up. The second derivative is positive, no matter where the ant walks inside the bowl.

CONCAVITY AND SLOPE

First Derivative	Second Derivative	Classification	Pictorial
$f'(x) < 0$	$f''(x) > 0$	Curve decreasing, concave up	Left side of $\smile$
$f'(x) > 0$	$f''(x) > 0$	Curve increasing, concave up	Right side of $\smile$
$f'(x) < 0$	$f''(x) < 0$	Curve decreasing, concave down	Right side of $\frown$
$f'(x) > 0$	$f''(x) < 0$	Curve increasing, concave down	Left side of $\frown$

Just as Fiona hands the card to Big Tony, two customers with long whiskers and M. C. Escher socks walk into my restaurant. One of them shouts, "I like concave up." The other shouts, "I like concave down." And then they leave.

Fiona shakes her head and says, "Get me out of this nut house."

EXERCISES ....

1. Describe the nature of the critical points for the function $y = x^3$. That is, describe all the points where the curve has a local minimum, local maximum, or inflection point.

2. Describe the nature of the critical points for $y = 3x^3 + 1$.

3. Find the critical points for $f(x) = 4 - 3x + x^2$.

4. Find the critical points for $f(x) = 5 - 6x + x^2$.

5. Find the local maxima, minima, and inflection points of

$$\frac{1}{3}x^3 + \frac{3}{2}x^2 + 2x + 1.$$

6. Find the local maxima, minima, and inflection points of

$$\frac{1}{3}x^3 + \frac{5}{2}x^2 + 6x.$$

7. Consider a curve defined by $x^3 - 9x - 2$. Where is this curve concave up and where is it concave down?

8. Consider a curve defined by $x^3 - 3x - 25$. Where is this curve concave up and where is it concave down?

9. For what value of k will

$$f(x) = x - \frac{k}{x}$$

have a relative maximum at $x = -4$?

10. For what value of k will

$$f(x) = x - \frac{k}{x}$$

have a relative maximum at $x = -1$?

7

Min-Max Pizza Applications

How many pizzas are consumed each year in the United States? How many words have you spoken in your life? . . . How many watermelons would fit inside the U.S. Capitol building?

—John Allen Paulos, *Innumeracy*

By now you can tell that my restaurant is my favorite hangout. Long ago, a portion of the building was first conceived as a priory and only later as a restaurant. The dining room floors are of random-width, golden-colored pegged pine. A few people play chess on seventeenth-century gaming tables in front of a fireplace.

Fiona takes an iron poker from the fireplace set, bends low to jab at the burning logs, adds a coarse chunk of red oak to the fire, and sets the poker back where it belongs. As she walks away, I see her run her fingers over one of my comfortable velvet couches, leaving faint tracks in the fabric.

"Fiona, today I want to do all sorts of fun problem solving with calculus. Now you'll know what calculus is good for. These are called *min-max* problems."

"Great!"

Big Tony and Rosario pull up chairs to hear my lesson.

B.C. MANSFIELD

FIGURE 12.
Luigi preparing
his garden.

"I want to grow tomatoes in my garden at home so that I can make homemade tomato sauce from scratch. I'll use the sauce for my pizzas" (Figure 12).

"Sounds delicious," says Fiona.

"Consider that I have 50 feet of fence that I want to attach to a wall of my house. I want the fence to surround my tomato garden to keep out animals. To make construction easy, I want the enclosure to be rectangular." I draw a figure showing the fence and the tomatoes (Figure 13).

"Sir," Rosario says, "I'll be right back with some pizza for us."

"Good," I say. Then I point to the figure.

"If one side of the fence has length x, then the fence available for the other two sides has a length of $50 - x$. Can you guess what the length of the remaining two sides is?"

Fiona puts up her hand. "The remaining two sides have a length of $(50 - x)/2$."

X

50-x
—
2

Tomatoes

50-x
—
2

FIGURE 13. Luigi's tomato garden enclosed by a fence 50 feet long.

Wall of Luigi's House

"Correct! Check your math by adding up all sides of the fence: $x + (50 - x)/2 + (50 - x)/2 = 50$. Now can you give me a formula for the area of the rectangular enclosure?"

Big Tony takes a gulp of his king-size root beer and blurts out, "The area is length times width or . . ." He writes on the tablecloth $A(x) = x (50 - x)/2 = \frac{1}{2}(50x - x^2)$.

"Correct, Big Tony. Next, I want you to notice that x can't assume just *any* value in this problem. It is constrained because we have only 50 feet of fence in total. In fact, the longest x could possibly be is 50, and the smallest it could possibly be is zero."

Big Tony has another swig of root beer as Rosario begins to serve us some artichoke and matzo-ball pizza.

"Here's our problem." I say. "I want to grow the most tomatoes that I can. This means I want to maximize the area of the garden. Yesterday we learned how to calculate local maxima of functions. We know the maxima occur at critical points where $f'(x) = 0$. For my garden, we want to maximize the function that describes the area of the garden. In our example, $A'(x) = \frac{1}{2}(50 - 2x)$, or $A'(x) = 25 - x$. This means that there is a critical point at $x = 25$."

"Is this a local maximum?" Big Tony says.

"Yes, because $A''(x) = -1$."

I look at Big Tony and Fiona and explain, "It seems that the most area we can obtain is 312.5 square feet."

"Why?" Fiona asks.

"At $x = 25$, the area is $\frac{1}{2}(50 \times 25 - 25^2)$, or 312.5 square feet. So, the dimensions that maximize the enclosure are 25 by 12.5 feet. But we should check to make sure there is not some value for area that is even larger than this local maximum. We should check the endpoints of our values for x, namely, $x = 0$ and $x = 50$."

"Wait, how could that be?" Fiona wonders. "We already found a local maximum corresponding to $x = 25$."

"Remember, yesterday I explained how a local maximum may not always be the absolute maximum over a given range for x. For example, take a look at this." I draw a picture on the wall that shows a local maximum, but one endpoint of the curve has a larger value than the local maximum (Figure 14). "Also remember that sometimes the local maxima or minima for a function may fall outside the range of x values in your problem, which, for our case, is $(0 \leq x \leq 50)$. It's always good to check."

I shake a little oregano and garlic on my pizza. Big Tony and Fiona do the same.

"It turns out," I say, "that at both $x = 0$ and $x = 50$ the area is zero. So we can rest easy that $x = 25$ produces the maximum area for my garden, which is 312.5 square feet."

"That's a good problem, Sir," Rosario says.

I nod and turn to Big Tony and Fiona. "Get some rest. Tomorrow we go to Florida for the next problem. Rosario, you stay in the restaurant to make sure it runs smoothly until we return."

"Yes, sir," Rosario says. "Your restaurant is in good hands."

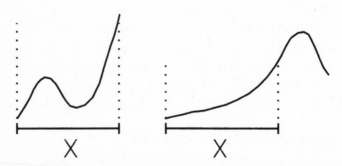

FIGURE 14. Ranges. (left) A local maximum is not necessarily the highest y value for a range of x values. (right) A local maximum may fall outside the range of x values of interest.

Fiona, Big Tony, and I are in a boat on the Florida Everglades. It is swampy and mosquitos are everywhere. It is rumored that piranha infest the warm waters.

"I just love the Everglades," I say. "The wildlife is awesome." I point to the right. "Look there, a purple gallinule walking across the lily pads." The bird has a purple head and breast. The beak is red on the tip. The legs are long and yellow.

I turn to Fiona and show her the map that I drew (Figure 15). "We are 4 miles from the shoreline. The world's tastiest pizza—pepperoni, onion, sausage, clam, anchovy, roasted purple gallinule, and poppy seeds—is 12 miles from the nearest point on the shoreline. Big Tony is hungry."

"You got that right," says Big Tony.

"It's slow going in the swampy water. Because the Everglades are full of marsh grass and lily pads, our boat travels at a measly 1 mile per hour. Once on land, the three of us can jog at 3 miles per hour. Big Tony is the limiting factor when it comes to our group walking quickly."

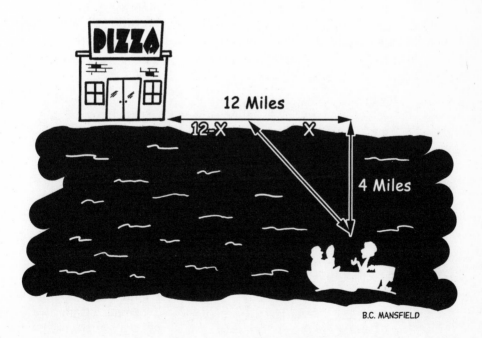

FIGURE 15. Luigi on a boat in the Everglades in search of pizza.

Big Tony thrusts his massive hand in my direction. "Are you insulting me?"

"I wouldn't think of it. Actually, I brought some scooters that could help us once on land."

Fiona puts her hands on her hips. "You brought us all the way to the Everglades to give us a min-max problem?"

"You bet, but you can also enjoy the wonderful surroundings. Look over there—a great blue heron."

"I have to admit it is pretty," Fiona says.

I turn to Big Tony. "Big Tony, toward what point on the shoreline should we point our boat in order to get to the pizza the fastest?"

"Luigi, do you mean we want to minimize the amount of time it takes us to get to the pizza?"

"Yes." I start sketching another diagram that shows our location relative to the pizza joint (Figure 16). "We are 4 miles from the shore. If we were to head directly toward the nearest point P on the shoreline, the pizza is 12 miles from this point."

"Twelve miles!" says Tony. "I'll die before we get that far."

"Don't worry, I brought snacks and soda. Now, as I was saying, the pizza joint is 12 miles from the point marked P. However, while we are in the boat, I think we should head toward the pizza somewhat, even though the slow boat has to travel a little farther in the water than it would travel if

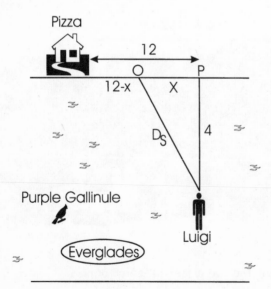

FIGURE 16.
Schematic diagram of Luigi's position.

we headed straight for the shoreline. We should aim for some optimum point on the land. Let's label this optimum shore point O. If x is the distance to be traveled from P to O, then $12 - x$ is the distance from O to the pizza."

"Fiona, what do we know about the triangle in the diagram?"

"By the Pythagorean theorem, we know that $4^2 + x^2 = D_s^{\,2}$, where D_s is the distance we row in the swamp. The square root of $(16 + x^2)$ is D_s. Thus, our journey would be one that goes to point O along the water and covers a distance D_s and then another distance $12 - x$ to the pizza. The total distance traveled to the pizza is the square root of $(16 + x^2)$ plus $12 - x$."

"Fiona, you're a genius! Now, according to elementary motion formulas, we know that speed is distance/time if we are traveling at a constant speed. Or, time equals distance/speed. The time we spend in the boat on the swamp is T_s hours $= D_s$ miles $/ 1$ mph $= \sqrt{16 + x^2}$ hours. So, the time we take to travel on land is $T_L = (12 - x)/3$. The total time to get to the pizza is . . ." I take Fiona's tan and teal lipstick and write on her arm

$$T(x) = \sqrt{16 + x^2} + (12 - x)/3 \text{ hours.}$$

"Notice that if we travel straight to point P on the shore and then to the pizza, we travel $4 + 12 = 16$ miles. In this scenario, $x = 0$, so the time it would take is $4 + 4 = 8$ hours. Big Tony's body is not in any condition to take such abuse. If we took our boat straight to the pizza, then $x = 12$, and it would take us this long to reach the pizza . . ." I write on Fiona's other arm

$$T(x) = \sqrt{16 + x^2} + (12 - 12)/3 \text{ hours.}$$

"This means it would take about 12.6 hours to get to the pizza if we took our boat directly to the pizza joint."

Big Tony looks nervous as he takes a deep breath. "Luigi, can we do better than 8 hours or 12.6 hours? Both are very long times."

"Let's find out. We want to minimize the time function. That means we first want to find the critical points of our time function."

"And that means," Fiona says, "that we want to find the derivative of the time function."

I sketch formulas on the floor of the boat:

$$T(x) = (16 + x^2)^{\frac{1}{2}} + (12 - x)/3 \text{ hours}$$

$$T'(x) = (1/2)(16 + x^2)^{-\frac{1}{2}}(2x) - 1/3 \text{ hours.}$$

"Fiona, when is the first derivative zero?"

She whips out a felt-tip pen from her pocket, takes my arm, and begins to draw on it.

"Fiona, here's a perfectly good pad of paper. No need to write on my arm."

"What? You draw on my arm and on the boat. Now you tell me you have paper?"

I shrug.

She sketches on the paper

$$(1/2)(16 + x^2)^{-\frac{1}{2}}(2x) - 1/3 = 0$$

$$x(16 + x^2)^{-\frac{1}{2}} - 1/3 = 0$$

$$\frac{x}{(16 + x^2)^{\frac{1}{2}}} - 1/3 = 0$$

$$\frac{x}{(16 + x^2)^{\frac{1}{2}}} = 1/3$$

$$3x = (16 + x^2)^{\frac{1}{2}}$$

$$9x^2 = (16 + x^2)$$

$$8x^2 = 16$$

$$x^2 = 2$$

$$x \approx 1.414 \text{ miles.}$$

"Excellent, Fiona. This would require an approximate time of . . ."

$$T(1.414) = (16 + 1.414^2)^{1/2} + (12 - 1.414)/3 = 7.771269 \text{ hours.}$$

"As I already told you, if we travel straight to the shore and then to the pizza, we travel 16 miles in 8 hours. If we take our boat straight to the pizza, then we need 12.6 hours. So, our 7.77-hour trip is a local minimum. Hopefully, Big Tony can survive the trip."

"What?" Big Tony says. "We did all this calculus to save a mere 8 – 7.77 = 0.23 hours, or about 14 minutes? It took us that long just to do the calculus! Luigi, sometimes you drive me nuts."

"Big Tony, I wanted to improve your mind. There are times when min-max calculus problems can solve really important problems. Also, I always wanted to show you and Fiona the amazing Everglades with its cypress, mangrove thickets, lush vegetation, and numerous islets."

A 6-foot-long alligator starts heading our way. I turn on the boat's engine and head toward shore. Hopefully, we will elude the alligator and have pizza in 7.77 hours. I look into Big Tony's large eyes as he flexes his arm muscles. He looks at me in a creepy way. I'm not sure what I fear most, Big Tony or the alligator.

EXERCISES........................

1. *Intervals.* Luigi discussed how it was often important to check the endpoints of intervals when trying to determine the absolute maximum for min-max problems. The following should make this more clear. Find the absolute maximum and minimum of $f(x) = 2x^3 - 4x^2 + 2$ on the closed interval $[-1, 1]$. (This means you want to find the absolute maximum and minimum for the range of x values $-1 \leq x \leq 1$.)

2. Following the example in problem 1, find the absolute maximum and minimum of $f(x) = 8x^3 - 16x^2 + 2$ on the closed interval $[-1, 1]$.

3. *The Pizza Tray.* Luigi wants to construct a rectangular pizza tray in which to make rectangular pizza snacks. To satisfy his customers, the bottom of the pizza must have an area of 60 cm^2. Luigi wants to minimize the amount of tray material used.

On the tray, Luigi needs an additional edge of 4 cm width on the right and left sides and edges of 3 cm on the top and bottom. (Figure 17 is a view of the pizza tray from above.)

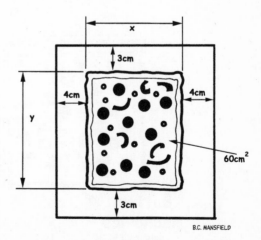

B.C. MANSFIELD

FIGURE 17. Luigi's rectangular pizza tray showing edges that allow him to grasp and carry the tray.

The edges allow him to grasp and carry the hot tray without getting his gloved hands stuck in the pizza. In the figure, how long should the pizza's horizontal dimension (x) be in order to minimize the amount of tray material used?

4. *Another Pizza Tray.* As in problem 3, Luigi wants to construct a rectangular pizza tray to make rectangular pizza snacks. This time, he wants the bottom of the pizza to have a larger area of 200 cm² to feed Big Tony. Given the same edge widths as in problem 3, how long should the pizza's horizontal dimension (x) be in order to minimize the amount of tray material used?

5. *The 3-D Pizza.* Luigi has invented a new form of pizza in the shape of a hollow box (Figure 18). Because the bottom surface is the most expensive with its imported Peruvian anchovies, the bottom costs $4 per square foot. The top has onions and is $1 per square foot. The four sides are pepperoni and cost $2 per square foot. Something for everyone at a party!

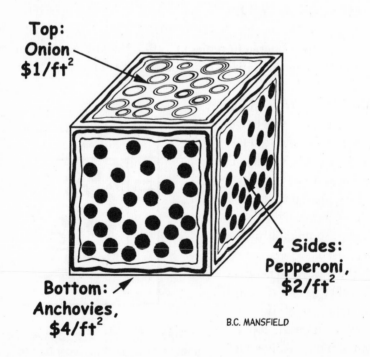

Top: Onion $1/ft²

4 Sides: Pepperoni, $2/ft²

Bottom: Anchovies, $4/ft²

B.C. MANSFIELD

. **FIGURE 18.** Luigi's 3-D pizza!

The box contains 25 cubic feet of space, which, on rare occasions, contains wonderful gifts for lucky customers. Customers find the gifts once they devour the pizza surfaces of the box. The base of the box is square. Your mission is to help Luigi find the dimensions that will minimize his cost.

6. Solve problem 5 given new numbers for the cost of each surface. The bottom now costs $6 per square foot, the top costs $3 per square foot, and the four sides are $4 per square foot.

7. *Sushi Pizza.* The price of Luigi's sushi pizza is $9 - 0.004x$ dollars, where x is the number of pizzas he makes each day. Notice how the price falls as he produces more pizzas, because people get stuffed and Luigi has to lower the price to motivate people to buy more. Of course, the pizzas cost Luigi a lot of money to make. In fact, the cost C of making and selling x pizzas is $4x + 1500$ dollars per day. How many pizzas should Luigi bake and sell every day in order to maximize his profit? (Luigi would like to buy Fiona a gift with the profit.)

8. How many sushi pizzas must Luigi sell each day to maximize his profit in problem 7 if the cost C of making and selling x pizzas is $3x + 1000$ dollars per day?

9. *Luigi's Number.* Luigi is staring at the wall at his favorite positive number x. The number is his favorite because when it is doubled, the number exceeds its square by the largest amount possible. For example, 0.1, when doubled, becomes 0.2, which exceeds its square (0.01) by 0.19. However, Luigi's number, when doubled, exceeds its square by more than 0.19. What is Luigi's number?

10. *Fiona's Number.* Fiona is staring at the wall at her favorite positive number x. The number is her favorite because the number exceeds its square by the largest amount possible. What is Fiona's number?

8

Exponentials and Logarithms

If a field ever needed to be brought out of mystery to reality, it is calculus. . . . Calculus is really exciting stuff, yet [the traditional course is] not presenting it as an exciting subject. . . . Calculus must become a pump instead of a filter in the pipeline.

—Robert White, American Mathematical Society Meeting

Pretty colors," Fiona says, looking at my latest artwork depicting the rolling Alban Hills south of Rome near the town of Frascati. This charming hill village has been my favorite summer residence for the last two years.

"Fiona, want to visit there one day with me?" I ask. "You'd love the world-famous white wine and ancient winding cobblestone streets." I turn on a bright light to better illuminate the painting that hangs in the private back patio section of the restaurant. All around the walls are shelves containing large jars of olives and peppers.

As Fiona wriggles her polka-dot fingernails near the glistening canvases, the artform casts a kaleidoscope of rosy reflections against the patio walls. She touches my arm as she looks at the quaint streets. "Psychedelic," she whispers. "Do you invite all your friends to the patio to see this?"

"Not all, Fiona."

There is a sudden shuffle of footsteps. "Sir," Rosario says, "I'm ready for today's calculus lesson." He comes ambling in with three glasses of ginger ale.

I look at Fiona for several more seconds, noting that she has tattooed the chain rule, which I taught her, on her ankle:

$$\heartsuit \ \frac{dy}{dx} = \frac{dy}{du} \times \frac{du}{dx} \ \heartsuit.$$

The tattoo appears to have a small heart on each side of the equation, but I not to look too closely so as not to appear overly curious.

"Rosario, today I'd like to start by reviewing exponentials and logarithms. When you see the term $\log_b x$, this can be thought of as the power you have to raise b to in order to produce x. For example, $\log_3 9 = 2$ because $3^2 = 9$. More generally, $b^y = x$ if $y = \log_b x$. Here are some other random examples." I write on the wall

$$\log_{10} 100 = 2 \text{ because } 10^2 = 100$$

$$\log_{10} 5 = 0.69897 \ldots \text{ because } 10^{\,0.69897\cdots} = 5$$

$$\log_{1/2} 16 = -4 \text{ because } (\tfrac{1}{2})^{-4} = 1/(\tfrac{1}{2})^4 = 16.$$

"Here are the rules for manipulating logarithms. The rules will come in handy during future soirees in the restaurant." I hand a card with a table outlining these rules to Fiona and Rosario.

RULES FOR MANIPULATING LOGARITHMS

Rule Name	Logarithm Rule
Product	$\log_b xy = \log_b x + \log_b y$
Quotient	$\log_b (x/y) = \log_b x - \log_b y$
Reciprocal	$\log_b (1/x) = -\log_b x$
Power	$\log_b x^p = p \log_b x$
Base change	$\log_c x = \log_b x / \log_b c$
Addition	None. Note: $\log_b (x + y) \neq \log_b x + \log_b y$

"Sir," says Rosario. "Can you give us some examples?"

"Sure. Here is an example that uses the product and power rules":

$$\log_2 4^3 6^7 = \log_2 4^3 + \log_2 6^7 \text{ (product rule)}$$
$$= 3\log_2 4 + 7\log_2 6 \text{ (power rule)}.$$

"Using the base change rule, we can change the base of 27 to the base of 3":

$$\log_{27} 1,000 = \frac{\log_3 1,000}{\log_3 27} = \frac{\log_3 10^3}{3} = \frac{3 \log_3 10}{3} = \log_3 10.$$

"Very often, mathematicians and physicists use the *common logarithm*, which is base 10 and is often written simply as log x, instead of $\log_{10} x$. Another frequently used logarithm is the *natural logarithm*, which is base e and is often written simply as ln x, instead of $\log_e x$. The number represented by e is Leonhard Euler's number. Euler lived during the 1700s. Like π, the digits of e go on and on without any obvious patterns. Also, like π, e appears in countless areas of mathematics. For example, in later lessons I'll tell you about the role e plays in the exponential growth of bacteria, money, and disease. Here is e expressed to high precision as . . ."

2.718281828459045235360287471352662497757247093699959574966 . . .

"Of course," I say, "for any conceivable practical application, you wouldn't need to use so many digits of e."

Fiona comes closer to examine the wonderful number, and I notice that she is wearing one of the latest pheromones—probably in a base of hentriacontane to make it particularly potent. "Luigi," she asks, "who was this Euler fellow?"

I walk over to the wall and insert a CD into a crack in the stucco surface. Suddenly the three of us hear *Mi fai venire il nervoso* sung by the renowned Italian tenor Luciano Pavarotti. The music emanates from carefully concealed speakers in the walls, ceilings, and flower pots. I slowly turn to Fiona, as if to heighten the suspense.

"For one thing, Euler is my hero. He was the most prolific mathematician in history. Even when he was completely blind, he made great contributions to modern analytic geometry, trigonometry, calculus, and number theory. Euler published over eight thousand books and papers, almost all in Latin, on every aspect of pure and applied mathematics, physics, and astronomy. In analysis, he studied infinite series and differential equations and introduced many new functions. His notations such as *e* and π are still used today. In mechanics, he studied the motion of rigid bodies in three dimensions, the construction and control of ships, and celestial mechanics. Leonhard Euler was so prolific that his papers were being published two centuries after his death. His collected works have been printed bit by bit since 1910 and will eventually occupy more than seventy-five large books."

Rosario raises his hand. "Sir, note that the number *e* can be defined as the sum of a series in which the series terms are the reciprocals of the factorial numbers: $e = 1/0! + 1/1! + 1/2! + \ldots = 2.7182818284590 \ldots$ (Recall that for a positive integer n, $n!$ is the product of all the positive integers less than or equal to n.) "

"Very impressive that you should know such a thing," I say. But now I feel I must outdo Rosario, so I write from memory the first "few" digits of *e* on the wall. After five minutes, I stop.

```
2.71828182845904523536028747135266249775724709369995957
49669676277240766303535475945713821785251664274274466391
93200305992181741359662904357290033429526059563073813232
28627943490763233829880753195251019011573834187 9
30702154089149934884167509244761460668082264800168477 41
18537423454424371075390777449920695517027618386062613 31
38458300075204493382656029760673711320070932870912744 37
47047230696977209310141692836819025515108657463 7
72111252389784425056953696770785449969967946864454905 98
79316368892300987931277361782154249992295763514822082 69
89519366803318252886939849646510582093923982948879332 03
62509443117301238197068416140397019837679320683 2
82376464804295311802328782509819455815301756717361332 06
98112509961818815930416903515988885193458072738667385 89
42287922849989208680582574927961048419844436346324496 84
87560233624827041978623209002160990235304369941 8
49146314093431738143640546253152096183690888707016768 39
64243781405927145635490613031072085103837505101157477 04
17189861068739696552126715468895703503540212340784981 93
34321068170121005627880235193033224745015853904 7
30419957777093503660416997329725088687696640355570716 22
68447162560798826517871341951246652010305921236677194 32
52786753985589448969709640975459185695638023637016211 20
47742722836489613422516445078182442352948636372 1
41740238893441247963574370263755294448337998016125492 27
85092577825620926226483262779333865664816277251640191 05
9004916449982893150566047258 0277
```

"Luigi!" says Fiona, "You're amazing!"

My heart skips a beat. I don't tell her that I used to impress a friend by writing random digits after the first few. After all, how would anyone know? However, my hobby is memorizing the digits of e, and all of these digits are correct.

I write an equation on the wall:

$$e = \lim_{n \to \infty} \left(1 + \frac{1}{n} \right)^n.$$

"Fiona, in my example this simply means that the larger the value for n that we use, the closer the approximation is for e. For example, if we calculate $(1 + 1/10)^{10}$, we get a value of 2.593742. If we calculate $(1 + 1/100)^{100}$, we get a value of 2.704814. If we calculate $(1 + 1/10{,}000)^{10{,}000}$, we get a value of 2.718146, and so forth."

I write another equation on the wall:

$$1 + e^{i\pi} = 0.$$

Rosario looks at the equation. "Tell us more."

"Some believe that this compact formula is surely proof of a Creator. Others have actually called $1 + e^{i\pi} = 0$ 'God's formula.' Edward Kasner and James Newman in *Mathematics and the Imagination* wrote, 'We can only reproduce the equation and not stop to inquire into its implications. It appeals equally to the mystic, the scientist, the mathematician.'"

"Luigi," Fiona says, "who first discovered this formula?"

"Leonhard Euler. The formula unites the five most important symbols of mathematics: 1, 0, π, e, and i (the square root of minus one). This union was regarded as a *mystic union* containing representatives from each branch of the mathematical tree: Arithmetic is represented by 0 and 1, algebra by the symbol i, geometry by π, and analysis by the transcendental e. The Harvard mathematician Benjamin Pierce said about the formula, 'That is surely true, it is absolutely paradoxical; we cannot understand it, and we don't know what it means, but we have proved it, and therefore we know it must be the truth.'"

"Interesting," says Fiona. "The formula is also so fundamental that it contains the three most basic operators (addition, multiplication, and exponentiation) as well as the two neutral units, or identity operators, 0 and 1 for these operations. This is the quintessence of mathematics."

As Fiona says the words *exponentiaton* and *quintessence,* I feel my heart begin to race. I've never heard Fiona use such technical and flowery language. The more technical she gets, the more my excitement escalates. "Evidently you two people have been studying," I say.

"We have!" says Rosario.

I turn to Fiona and say, "I love e because the function e^x is its own derivative. In other words, the value of y in $y = e^x$ is always the same as the rate of change of this function. At each point in the graph of $y = e^x$, the value for y and the slope are equal." I draw a table on the wall, the first row of which shows that the derivative of e^x is e^x itself.

EXPONENTIAL RULES

Rule Name for Derivative	Exponential Rule
Exponential	$(e^x)' = e^x$
Exponential (chain)	$(e^u)' = e^u u'$
Exponential base b	$(b^x)' = (\ln b)b^x$
Exponential base b (chain)	$(b^u)' = (\ln b)b^u u'$
Logarithm	$(\ln x)' = 1/x$
Logarithm (chain)	$(\ln u)' = (1/u)u'$
Logarithm base b	$(\log_b x)' = (1/\ln b)(1/x)$
Logarithm base b (chain)	$(\log_b u)' = (1/\ln b)(1/u)u'$

"Here are some examples. The first one shows the exponential chain rule where the derivative is $e^u u'$. In this case, $u = 2x^4$ and $du/dx = 8x^3$." I write on the wall

$$(e^{2x^4})' = 8x^3 e^{2x^4}.$$

"Here is an example using the exponential rule base b and the accompanying chain rule":

$$(3^x)' = (\ln 3) \times 3^x$$

$$(3^{x^2})' = (\ln 3) \times 2x \times 3^{x^2} \quad (u = x^2)$$

"Here is an example using the logarithm chain rule. Given $y = \ln(8x^2 - 2)$, we find that . . ."

$$y' = \frac{1}{8x^2 - 2}(16x) = \frac{16x}{8x^2 - 2} \; (u = 8x^2 - 2).$$

I turn to Fiona. "That's enough for today. Our lesson focused on exponentials and logarithms—two important areas of mathematics."

The Pavarotti song fades from the loudspeakers. It is finished. My heart rate returns to normal. "I suggest we rest now. Tomorrow we have a big day with limits and continuity."

Fiona seems deep in contemplation. After some time she says, "Your restaurant is so wonderful, so spacious. I grew up in a miserable one-room shack in Novosibirsk, Russia. We had no electricity or running water. Ugh. I hated it. Sometimes tourists came by and took photographs." Fiona clenches her fists and then looks away from me. "I felt like I was in a zoo. My mother washed clothes by hand in an abandoned tub using water from a nearby gas station."

I am startled by her sudden and personal outburst. "But I thought your father was a famous professor?"

"Yes," she says softly, "but that was much later. He blossomed later in life. When I was growing up, we were dirt poor."

Fiona looks dreamily into the huge jars of olives on the shelves, and then she shrieks. "What's in there?"

"Nothing to worry about," I say. "Some of the jars without olives are actually small saltwater aquaria." I smile as I follow her gaze. Inside the jars, connected by transparent pipes, is an oasis. Life is everywhere: beautiful naked gastropods, organisms that resemble land snails, sea snails, limpets, and whelks. The creatures remind me of the time I spent exploring the seaports of the southeastern end of Italy along the Gulf of Taranto.

I turn to Fiona. "Care for some calamari?"

"It's late," she says.

After Fiona and Rosario leave, I gaze out the ceiling skylight. Windows in the nearby buildings form patterns of crystalline checkerboards that scintillate in the dying sunlight. Some of the glass reflects indigo beams of light into the heavens. I look toward the sky, always a source of pleasure, with cotton candy clouds and gaseous ivory mists.

EXERCISES........................

Evaluate the following:

1. $\left(2e^{\sqrt{x}}\right)'$

2. $\left(\pi e^{\sqrt{x}}\right)'$

3. $\left(3^{x^2}\right)'$

4. $\left(4^{x^3}\right)'$

5. $\left[\ln\left(8x^2+2\right)\right]'$

6. $\left[\ln\left(10x^2-5\right)\right]'$

7. $\left(-e^{\sin x}\right)'$

8. $\left(-e^{-2\sin x}\right)'$

9. $\left[\ln\left(e^x+x\right)\right]'$

10. $\left[\pi\ln\left(e^x+x^2\right)\right]'$

Limits and Continuity

uigi," Fiona calls to me. "Why is everyone looking so intently at tonight's menu?"

I hand the first page of the new menu to Fiona. "Here, take a look."

"Holy mackerel!" Fiona exclaims. "You're asking people to figure out calculus problems to determine the price of dishes?"

"Yes, it will improve their brains. Studying calculus is for the mind as Kung Fu is for the body. By the way, notice the notation d over dx. I want to remind you that it means the derivative with respect to x." I write on the restaurant wall

$$\frac{d}{dx}(x^2 + 5) = (x^2 + 5)'.$$

LUIGI'S ITALIAN MENU

Entree	Price
Eggplant Parmigiana Fresh sliced eggplant, breaded and fried until crisp, topped with cheese and tomato sauce.	$\$\dfrac{d}{dx}(x^2+5)$ at $x=5$
Fiona's Special Angel-hair pasta with fresh scallops, shrimp, and chopped roma tomatoes sautéed in extra-virgin olive oil and garlic, perfectly seasoned. Luigi adds plump Gulf shrimp and tender, juicy scallops. Tossed. That's Italian!	$\$\dfrac{d}{dx}(x^2+5)$ at $x=3$
Strüdel Val Gardena Delicious mixed vegetables steam cooked and quickly sautéed in fresh butter, then rolled in a thin layer of fresh hand-made egg pasta.	$\$30\displaystyle\int_{-1}^{1}x^2dx$
Tiramisu alla Fiona Cake soaked in rum and espresso coffee, topped with zabaglione and cream-cheese sauce, sprinkled with chocolate and espresso. Mamma Mia!	$\$\dfrac{d}{dx}(\pi x+5)$
Calculus Salad Lettuce, tomato, onion, bell pepper, mushrooms, artichoke hearts, and chunks of mozzarella cheese in the shape of π and integral symbols.	$\$\pi^{\pi}$

"Also notice the integral symbol $\int$. I haven't taught you about that yet."

"Luigi, is there no limit to your strange behavior? I mean, I like your quirks, but don't you think the customers will find it disturbing?"

"Not at all," a customer yells from the back of the room. "We love calculus and pizza."

I grin. "Fiona, could you repeat exactly what you said? But please say the word *limit* louder."

Fiona thinks about it. "Luigi, is there no *limit* to your strange behavior?"

"You said limit. And that reminds me. Today we want to talk about limits."

I motion for Fiona to sit down. "Fiona, when we say we want to take the limit of a function, it means we want to find out the function's behavior as we approach a particular value, for example, we might want to know the value of the function $f(x)$ as x approaches a number, a."

Big Tony sits down with us, and I continue. "Sometimes we can get a good idea about the limit of a function by examining its graph. We can also compute limits without graphing a function. Consider the simple parabola, $y = x^2$. As x approaches zero from either the positive or negative direction, the value of the function $f(x)$ approaches the value of zero. For a simple case like this, the limit of the function as x approaches some number a is just a^2. You just insert a into the function for x. We can write this as follows." I write on the tablecloth

$$\lim_{x \to 0} x^2 = 0.$$

I turn to Big Tony. "Big Tony, solve this":

$$\lim_{x \to 0}(x^2 - 3) = ?$$

Tony waves his mozzarella calzone at me. "Luigi, that's easy. Simply insert the value zero for x, and you find the limit is -3."

"Very good. Simple insertion often works if we don't produce a number that has zero in a denominator or a negative number inside a square root. But sometimes limits don't exist. For example, consider $y = 1/x^2$. As x approaches zero, y approaches infinity, and we usually say that the limit does not exist."

I turn to Fiona. "Let us try a more complicated example. Consider this":

$$\lim_{x \to 2} \frac{x - 2}{x^2 - 4}.$$

Fiona looks at me. "Luigi, when x is 2, the denominator is zero. The limit of this function must be undefined."

"Not so. Don't give up so easily. The limit can exist for some fractions with zero as the denominator. First try to simplify the fraction":

$$\lim_{x \to 2} \frac{x - 2}{x^2 - 4} = \lim_{x \to 2} \frac{x - 2}{(x + 2)(x - 2)} = \lim_{x \to 2} \frac{1}{x + 2} = 1/4.$$

"Fiona, you were right to notice that the *function* $(x - 2)/(x^2 - 4)$ is not defined at $x = 2$. It has a 'hole' at this point. At other points, the curve just looks like $1/(x + 2)$. Remember, when you evaluate a limit, you are only concerned about values of a function *near* a particular point and not *at* the actual point. One strange limit is the following":

$$\lim_{x \to +\infty} \frac{\sin x}{x} = 1.$$

"Wow," says Fiona. "When $x = 0$, we seem to have the fraction 0/0."

"One way to look at this is that the value of x in radians is very close to the value of sin x for small values of x. For example, $\sin(0.003) \approx 0.002999996$. As you use values of x closer and closer to zero, the two functions hug and approximate each other—a marriage made in heaven." I pause and look at Big Tony. "But, in general, if you substitute a value a for x in a function and the resultant fraction is 0/0, you know *nothing* about the limit. You have to do more work to determine the limit."

Fiona says, "Let me see if I remember correctly. A radian is a measure of an angle, and one radian is $180/\pi$ degrees?"

"Correct."

Big Tony seems to be distracted by a Mona Lisa poster that Rosario has placed on the wall. Beneath the poster are the words:

$5 OFF ANY ENTREE
IF YOU KNOW
THE DERIVATIVE
OF e^{x^2}.

"Big Tony, in this age of modern computers, we can estimate the limit of many functions by trying numbers closer and closer to the limit value. For example, we might try a BASIC program like this one." I pull a card from my pocket and hand it to Fiona, who looks at it and then passes it to Big Tony.

BASIC Program for Probing Limits

```
10 REM Limit of a Function
15 REM If you don't use BASIC, your mission is to rewrite
17 REM this in your favorite language—like C++ or Java
20 DEF FNY(X) = (x-2)/((x*x)-4)
25 REM The value at which to test the limit:
30 A = 2
35 REM The number of times to test on each side of the limit:
40 K = 5
45 REM Print the column headers for the output table
50 PRINT "A - H", "FNY(A - H)", "A + H", "FNY(A + H)"
55 REM Test the function at points close to A:
60 FOR N = 1 TO K
70      LET H = 1/10^N
80      PRINT A-H, FNY(A-H), A+H, FHY(A+H)
90 NEXT N
100 END
```

"If you don't like BASIC, rewrite this in your favorite language—like Pascal, C++, or Java. Look at my code. In step 20, we define the function of interest, in this case $(x - 2)/(x^2 - 4)$. In step 30, we assign the value for which we want the limit, in this case 2. Step 60 repeats the computation five times. The purpose of step 70 is to produce successively smaller values; in this case, H equals 0.1, 0.01, 0.001, and so forth. These small numbers can be used for computing the function's value closer and closer to $A = 2$. Think of this process as a lion slowly creeping up on its prey."

Fiona nods. "And if the prey escapes, perhaps that's like a limit that is not defined at a point."

"I'll have to think about that one." I continue. "The results are printed in step 80. Of course, this is not a perfect method, because the function might suddenly change behavior, for example, 0.000000000001 closer to A than the program has checked; nevertheless, the program is a great way of guessing the limit for many functions you will encounter." I pull out a pocket calculator from behind a potted plant. Fiona seems suitably impressed. I look her directly in the eyes. "Let's run the program on my calculator for our example. Here's the function and computer output . . ."

$$\lim_{x \to 2} \frac{x-2}{x^2-4}.$$

BASIC PROGRAM OUTPUT

$A - H$	$\lim_{x \to 2} \dfrac{x-2}{x^2-4}$, $x = A - H$	$A + H$	$\lim_{x \to 2} \dfrac{x-2}{x^2-4}$, $x = A + H$
0.9	0.256410256	2.1	0.243902439
1.99	0.250626566	2.01	0.249376559
1.999	0.250062516	2.001	0.249937516
1.9999	0.25000625	2.0001	0.24999375
1.99999	0.25	2.00001	0.25

"Notice as we approach 2 from below or above, the values for the limit appear to converge to 0.25, which is what we expect."

"Luigi," Fiona says, "let's go to the back patio. The weather is so nice."

"Certainly."

When we enter the patio, we notice a blue butterfly rising from a man-made trickling stream. It drifts toward Fiona, moving like a feather floating in the breeze. Her hand goes to her face as her eyes open wide in wonder. The butterfly, floating at the height of her head, flashes various shades of blue and crimson in the bright afternoon sunlight. At the edges of the membranous wings are little sparkles, as if the light were igniting tiny dust particles in the air.

"It's so beautiful," Fiona says as the butterfly drifts away into the distance.

I nod. "Let's try to evaluate another kind of limit, one in which we observe the behavior of a function as x approaches infinity." I write on a napkin

$$\lim_{x \to +\infty} \frac{5x^7+1}{2x^5+6}.$$

"Here both the numerator and denominator approach infinity. One good strategy to determine how the fraction behaves in the limit is to divide the numerator and denominator by x^5, the highest power of x in the denominator":

$$\lim_{x \to +\infty} \frac{5x^7 + 1}{2x^5 + 6} = \lim_{x \to +\infty} \frac{5x^2 + 1/x^5}{2 + 6/x^5}.$$

"Next we can take the limit of each term":

$$\frac{\lim_{x \to +\infty} (5x^2) + \lim_{x \to +\infty} (1/x^5)}{\lim_{x \to +\infty} 2 + \lim_{x \to +\infty} (6/x^5)} = \frac{\lim_{x \to +\infty} (5x^2) + 0}{2 + 0} = \frac{\lim_{x \to +\infty} (5x^2)}{2} = +\infty.$$

Big Tony puts his big fist on the table. "So the limit is really positive infinity even though it looked like the limit was ∞/∞ before you simplified?"

"That's right." I pause. "My interest in limits brings me to my favorite rule in all of calculus. It is called L'Hôpital's rule—an amazing formula probably due to the Swiss mathematician Johann Bernoulli (1667–1748) who was the teacher of the French nobleman Marqius of L'Hôpital (1661–1704). Consider any fraction, such as our previous friend . . ."

$$\frac{\sin x}{x}.$$

"If both the numerator and the denominator approach infinity or zero for some limit, we call the limit an *indeterminate form*. In our example, sin x and x both approach zero as x approaches zero. We can apply L'Hôpital's rule to any indeterminate form."

L'Hôpital's Rule

$$\lim_{x \to a} \frac{f(x)}{g(x)} = \lim_{x \to a} \frac{f'(x)}{g'(x)}$$

Fiona draws a box around the rule and looks at me. "Let me see if I understand. If I have a limit that looks like this . . ."

$$\frac{0}{0} \text{ or } \frac{\infty}{\infty},$$

"then I can examine the derivative of the numerator and denominator to help me find the actual limit, which I'm guessing might happen to be 0, 1, or ∞."

"You got it. We can try L'Hôpital's rule for this equation as x approaches zero":

$$\lim_{x \to 0} \frac{\sin x}{x} = \lim_{x \to 0} \frac{(\sin x)'}{x'} = \lim_{x \to 0} \frac{\cos x}{1} = \frac{1}{1} = 1.$$

"Superb," Fiona says.

"Let's try another. What is the limit of the following?"

$$\lim_{x \to \infty} \frac{e^x}{x}.$$

"Step one," I say. "First, determine if this is an indeterminate form."

"Yes, it is," Fiona says. "The fraction is ∞/∞ in the limit."

"Step two. Take the derivative of the numerator and the denominator to find the limit":

$$\lim_{x \to \infty} \frac{e^x}{x} = \lim_{x \to \infty} \frac{(e^x)'}{x'} = \lim_{x \to \infty} \frac{(e^x)}{1} = \infty.$$

"So, infinity is the answer." I pause. "Sometimes we need to use L'Hôpital's rule several times to find an answer. Consider the following":

$$\lim_{x \to 0} \frac{1 - \cos x}{x^2} = \lim_{x \to 0} \frac{\sin x}{2x} = \lim_{x \to 0} \frac{\cos x}{2} = \frac{1}{2}.$$

"Notice how I took the derivative once, and then I took the derivative again to further simplify the expression. In the end, I was able to show that the limit was 1/2 as x approaches zero." I pause. "Fiona, there are some other assumptions one has to consider when using L'Hôpital's rule, such as that it can only be applied if $\lim_{x \to 0} f'(x)/g'(x)$ exists —but my introduction is enough to get you started on some basic problems."

I reach into Fiona's left pocket.

"What do you think you're doing?" she asks.

I hand Fiona her maroon sparkle lipstick made from a biodegradable soy product. "Fiona, I want you to doodle on my arm. Make a wacky, wavy shape."

Fiona draws a flowing curve until I tell her to stop.

"Fiona, your curve is *continuous*. It is continuous because you never lifted the lipstick from my arm as you drew the curve. But let me give you a more rigorous definition of continuity. A function $f(x)$ is continuous at a point $x = a$ if . . ." I draw on her arm the definition for continuity.

Continuity at *a*

---

1. $f(a)$ is defined

2. $\lim_{x \to a} f(x)$ exists

3. $\lim_{x \to a} f(x) = f(a)$

"An example of a function that does not meet condition 1 occurs when the function is not defined at a point *a*, for example, when *a* is a negative number and the function is $\sqrt{x}$. We can't talk about continuity of the function at this point if the function isn't even defined for negative numbers."

"What does condition 2 mean?" Big Tony asks.

"Condition 2 simply means that the function must be tending toward a particular value as *x* approaches *a*. Condition 3 indicates that in the limit the function is approaching the value of the function at *a*."

"Give us some examples of condition 3," says Fiona.

"Here are two examples." I draw a curve (Figure 19). "This curve has two discontinuities. The discontinuity on the left shows a single point that has been displaced upward. The function is defined at this point. It does have a value. The limit does exist at a_1. However, the limit at this point and

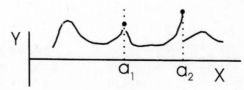

FIGURE 19. A curve with two discontinuities.

the value of the function at
this point are different. The
function fails condition 3, so
the curve is not continuous at
a_1. In the discontinuity at the
right, the limit doesn't even
exist, because as we approach
$x = a_2$ from the left, we find a
different limit than we find as

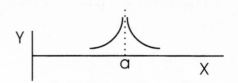

FIGURE 20. A curve with one
discontinuity.

we approach from the right. At a_2, the function fails condition 2."

I draw another figure (Figure 20). "This figure shows a curve that fails
all three tests because $f(a)$ is not even defined at $x = a$. The curve just sky-
rockets to infinity at this point, and we say the curve has no limit at this
point."

Fiona appears deep in thought. "Luigi, can a curve be continuous if it
has points at which the derivative is not defined?"

"Yes," I say. "And some of these kinds of curves can be quite fascinating.
Curves can be continuous even if they have points at which they are not dif-
ferentiable. For example, $y = |x|$ is continuous everywhere—you can draw
it without lifting your pen—even though the derivative is not defined at
$x = 0$ where a sharp corner exists." I draw a plot of $y = |x|$ (Figure 21).

"Why do you say it doesn't have a derivative?" Fiona asks.

"Do you see the sharp corner $x = 0$? A function doesn't have a definite
tangent at a corner like this. Think of a seesaw. Think of the long plank
of wood atop the pivot point of a seesaw. Further visualize the seesaw
teetering back and forth because there is no single tangent to a corner on
which it balances. Similarly, a function is not differentiable at a point in
which the derivative has no
particular value. I often think
about curves as being differentiable everywhere only if
they are smooth and have no
sharp edges or jumps. On the
other hand, my favorite curve
that is nondifferentiable everywhere (because it is so pointy!)
is the Koch curve." I bring out

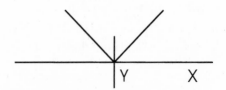

FIGURE 21. The curve $y = |x|$ is
continuous even though the derivative is undefined at $x = 0$.

a card from my pocket showing how to make a Koch curve. (Figure 22).

"Its edge looks a little like a snowflake's edge," Fiona says.

"You can construct an edge of the Koch curve by continually replacing the center third of every straight-line segment with a V-shaped wedge. After numerous generations, the length of the curve would become so great that you could not carefully trace the path in your entire lifetime."

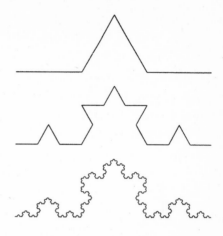

FIGURE 22. Creating a Koch curve.

Big Tony looks at the Koch curve. "Did someone recently discover the Koch curve?"

"In 1904, the Swedish mathematician Helge von Koch proposed this curve. It's considered to be a fractal curve because it displays similar structures as you magnify the crinkly edges. You can't draw tangent lines at any points, so the curve is nondifferentiable at every point."

"What's a fractal?" Tony asks.

"Generally speaking, a fractal is an object or pattern that exhibits similar structures at different-size scales. Think of the leaves of a fern, the branchings of blood vessels, or the edges of a coastline."

"Wild," says Fiona. "It's strange to imagine a fractal shape like the Koch curve with infinitely many bumps between bigger bumps, in a finite region in space, and realize that it is still a continuous curve."

"Yes, mathematics can be strange." I press a button under the table and organ music pours from concealed speakers in the legs of the table. It is Johann Sebastian Bach's *Toccata and Fugue in D Minor*. "Intuitively you might imagine that this curve, which contains infinitely many edges in a finite space, should begin to break up or become discontinuous. Continuity for a function, f, may also be informally defined as follows: points that are very close together are mapped by f into points that are very close together. Sometimes this is hard to visualize intuitively." I pause. "Consider the following function":

$$x \sin\left(\frac{1}{x}\right).$$

"The graph of this function for |x| < 1 shows smaller yet more rapid oscillations as it approaches zero. The limit of this function as x approaches zero is zero." I take out a card showing the graph of this function (Figure 23).

"Fiona, we can define a very similar function . . ."

$$\psi(x) = x \sin\left(\frac{1}{x}\right)$$

when

$$x \neq 0;$$

otherwise,

$$\psi(x) = 0.$$

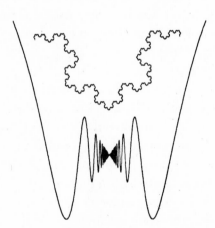

FIGURE 23. $\psi(x)$

This mathematical function has the property of being continuous everywhere, even though it has infinitely many oscillations in the neighborhood of x = 0 where the size of the oscillations becomes infinitely small. The frequency at x = 0 is infinite and the spacing between the maxima is zero! A Koch curve (top) is superimposed for comparison.

"Like a Koch curve," Fiona says, "this curve has infinitely many bumps, decreasing in size, in a finite region of space."

"Is $\psi(x)$ continuous?" Big Tony asks.

I look outside at the waning light. It is raining, cool and light, and the air smells fresh. I can just see the tall skyscrapers a mile away under a floodlit glare.

"Fiona, because sin x is everywhere continuous, and $1/x$ is continuous for $x \neq 0$, it follows that the composite function $\sin(1/x)$ is continuous for $x \neq 0$. Therefore, $\psi(x)$ is a simple mathematical function that is everywhere continuous, even though it has infinitely many oscillations in the neighborhood of $x = 0$, where the size of the oscillations becomes infinitely small. However, unlike the fractal Koch curve, $\psi(x)$ is differentiable (i.e., smooth) for $x \neq 0$, even though the frequency near $x = 0$ approaches infinity and the spacing between the maxima approaches zero." I take a breath. "Don't you love mathematics? It really lets the soul soar."

Fiona repeats the word *soul* very slowly as if imitating a process approaching a limit. I feel quiet and relaxed. Big Tony begins to chant *Sei Scemo*, a soulful tune I recall my mother singing to me as a child. The song's ending seems to be slowing. Is Big Tony slowing intentionally?

A fly alights on Big Tony's head and then departs. No one moves. The fly seems suspended, motionless. Reflections from Big Tony's shiny skin and wine glass make me feel that I am next to the periphery of some gigantic crystal as my image is many times reflected.

I shake my head, and the world is back to normal. Perhaps I am just nervous about my plan to teach Fiona about integrals in the next few days. There are dangers. Perhaps I have gone too far with Fiona. But what is too far when I feel I've made a fool of myself? Perhaps my subliminal attraction to Fiona has become too obvious to her.

"Luigi," Fiona says. "It is getting late."

"You're right. Let's continue tomorrow."

I wave good-bye as Big Tony and Fiona walk down Fifth Avenue to their separate apartments. It's time to prepare for tomorrow's lecture on related rates. I feel a peculiar sense of déjà vu. . . .

A few minutes later, I am in my apartment above the restaurant. It's been a long day. I lie in bed listening to the sounds of crickets coming in through the window, so close they sound as if my room has been magically transported to the countryside. I even smell flowers—a fruity aroma that seems out of place in the city. I am drifting off to sleep, and as I do, I whisper the words *calculus* and *freedom* and dream about derivatives and continuity—and about my preparations to someday teach Fiona about integrals while we sail on a shoreless sea.

EXERCISES.........................

Evaluate the following limits:

1. $\lim\limits_{x \to 6} \dfrac{x^2 - 36}{x - 6}$

2. $\lim\limits_{x \to 7} \dfrac{x^2 - 49}{x - 7}$

3. $\lim\limits_{x \to \infty} \dfrac{5x^2 - 3x + 2}{2x^2 - 7x + 3}$

4. $\lim\limits_{x \to \infty} \dfrac{7x^2 - 2x + 1}{3x^2 + 5x}$

5. $\lim\limits_{x \to 0} \dfrac{2x}{\sqrt{x+1} - 1}$

6. $\lim\limits_{x \to 0} \dfrac{\pi x}{\sqrt{x+1} - 1}$

Use L'Hôpital's rule to find the following limits:

7. $\lim\limits_{x \to 0} \dfrac{1 - e^x}{2x}$

8. $\lim\limits_{x \to 0} \dfrac{1 - e^x}{3x}$

9. $\displaystyle\lim_{x \to +\infty} \frac{2 \ln x}{x}$

10. $\displaystyle\lim_{x \to +\infty} \frac{\ln x}{2x}$

10

Related Rates

Imagine that you're dressed in gossamer. You have delicate white wings and are sitting on a fluffy cloud. You are experiencing the greatest dessert ecstasy of your life. You are in Heaven, and Heaven is in your mouth. That's Tiramisu!"

—Craig Miyamoto, "What Is Tiramisu?"

The scope of calculus is oceanic.

—Jan Gullberg, *Mathematics: From the Birth of Numbers*

L uigi," says Fiona, "someday I want to get a Ph.D. in mathematics so that I will know a lot of calculus."

"Great," I say with a smile. "I don't mean to brag, but I, in fact, have three such Ph.Ds."

She punches me in the arm. "What do you mean? Three!"

"Yes, and I must point out to you that in the next few days I won't have time to teach you *all* of calculus. Calculus is an endless oasis, and I am only giving you a few sips of its refreshing waters. For example, I won't be teaching you much if anything about such exotic and wondrous topics as differential equations, Fourier series, vector fields, Lagrange multipliers, Green's theorem, partial derivatives, gradients, Maclaurin series, divergences, curls, surface integrals, gamma and beta functions, elliptic integrals, Jacobians—"

"Oh my God. Stop! You don't have to try to impress me."

"I was only meaning—"

"Stop. You *have* impressed me, and you have also convinced me. Someday, I want to learn all the calculus I can. I will never stop learning. I want my brain to grow and grow and grow. I want to transcend space. Transcend time. I want to plot the trajectory of angels and calculate the tangent of heaven."

For a second, I am silent. She is quite the poet! And she is so intelligent that in a few years she will surpass me. But is Fiona trying to impress me now? Is she serious or is she mocking me?

I take a deep breath. "Fiona, today we will spend time on related rates problems."

"Tell me more."

"Some calculus problems require us to find the rate of change of two or more variables that are related to another common variable. To solve these problems, we often use implicit differentiation with respect to time to solve for a rate of change."

Rosario comes in carrying several gallon cans of imported oregano. He looks from me to Fiona. Perhaps he is confused by the words *implicit differentiation.*

"Hi, Rosario," I say. "Join us. A few days ago, we studied implicit differentiation. This will all become clear as I give an example. Today we are interested in the rates of change of two or more related variables."

The three of us sit down and order tea and tiramisu. The tiramisu is my special recipe that calls for eggs, mascarpone cheese, ladyfingers, cream, nutmeg, espresso coffee, brandy, marsala, rum, and a little bit of sugar, cocoa, and shaved chocolate.

"Rosario, consider a piece of pepperoni that is 6 feet in diameter."

"*Mamma mia*, that's a lot of heartburn," says Rosario.

"Yes. Now visualize this. It's evening, and the pepperoni rolls toward a street lamp that is atop a 14-foot-tall post. The pepperoni is traveling toward the lamp at a rate of 4 feet per second. The pepperoni casts a shadow on the ground. I'm interested in the distance between the shadow's edge farthest from the lamppost to the point of contact between the pepperoni and the ground—in other words, the length that approximates the pepperoni wheel's shadow. At what rate is this length changing when the point of contact is 10 feet from the lamppost?"

I begin to draw a diagram on the restaurant wall (Figure 24). "Let's

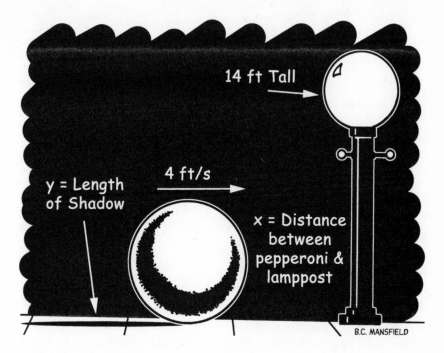

14 ft Tall

4 ft/s

y = Length
of Shadow

x = Distance
between
pepperoni &
lamppost

B.C. MANSFIELD

FIGURE 24. Tracking the shadow of a rolling pepperoni slice.

think about what we know so far and what we don't know as we label the graph. Let's label the distance from the pepperoni to the lamppost as x. Let's use y to label the distance between the shadow's edge and the pepperoni's point of contact with the ground."

Fiona looks closely at the diagram. "I see two right triangles in your drawing. One has a leg length of y, which represents the shadow length, and it has another leg that is 6 feet tall. A larger triangle has a base of $x + y$. This is the triangle with the lamppost as one of its sides. These two triangles are called *similar* triangles because all their corresponding angles are the same. The ratios of the sides of similar triangles are equal, so this means . . ." She sketches an illustration (Figure 25) and writes

$$\frac{6}{y} = \frac{14}{x+y}.$$

"Or we can write this as . . ."

$$6(x + y) = 14y \text{ or } 6x = 8y$$

or

$$3x = 4y.$$

I nod. "We want to know the rate of change of the length of the shadow when the pepperoni is 10 feet from the lamppost, that is, when $x = 10$."

Rosario raises his hand. "Sir, the rate of change of the shadow's length is the derivative of y with respect to time."

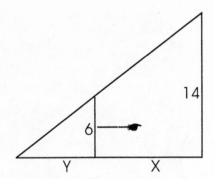

FIGURE 25. Fiona's schematic diagram.

"Yes! We want to know dy/dt when $x = 10$. Because the pepperoni is moving toward the lamppost at 4 feet per second, we now that $dx/dt = -4$. The derivative is negative because x is getting smaller as measured from the lamppost as time progresses."

"Luigi," Fiona says, "let me try to solve this. We know the relation between x and y is $3x = 4y$. We can differentiate this equation implicitly, as you explained the other day." She writes

$$3\frac{dx}{dt} = 4\frac{dy}{dt}.$$

"We know that dx/dt is -4, so $3(-4) = 4(dy/dt)$, which means that dy/dt is -3. When the pepperoni is 10 feet from the base of the lamppost, its shadow is shortening at a rate of 3 feet per second."

I stand up. "Fiona, you are absolutely correct! I should point out that our solution is an approximation. What we actually computed was the rate of change of y in this diagram." I draw on the restaurant wall (Figure 26). "You can see that the actual shadow length is slightly larger than y because the light rays will be tangent to the pepperoni and will not really go through the

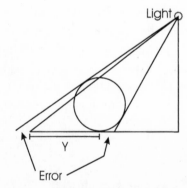

FIGURE 26. Slight errors in our approximation of shadow lengths.

top point of the pepperoni. Look carefully at my diagram. Also, there will be a tiny bit of shadow to the right of the contact point with the ground. But such is life. Approximations are fine to get the point across."

Finally, our tiramisu arrives in little glass cups that sparkle like diamonds. The dessert looks delicious.

Rosario pierces the surface of the tiramisu with his spoon. "Fiona," he says, "did you know that tiramisu was the favorite of Venice's courtesans, whenever they needed a stimulant or a 'pick-me-up,' which is the literal translation of tiramisu? The Venicians needed the coffee's caffeine to strengthen themselves between their romantic encounters."

"Rosario," I say, "there's a lady present. Please refrain from talking about romantic encounters, or at least wait until we are done with our lessons."

"Yes, sir."

I reach under the table and withdraw a meatball. I place it in on the table in front of Fiona and Rosario. They stare at the meatball and then at me.

"Rosario, consider one of my meatballs growing in volume at a rate of 1 cubic centimeter per hour."

"Sir, that's crazy. Meatballs don't grow."

"Rosario, use your imagination. Pretend this is an alien meatball that grows" (Figure 27).

FIGURE 27.
An alien meatball
growing.

B.C. MANSFIELD

"Or imagine fungus growing on the meatball," Fiona says. And then: "Gross!"

I nod. "Or imagine the meatball rolling down a hill and growing as it picks up debris. How fast is the meatball's *surface area* growing at the moment when the radius of the meatball is 10 cm?" I draw on the wall (Figure 28).

Rosario speaks with his mouth full of tiramisu. "What a fascinating question, sir."

"To find the answer, consider that the surface area of a sphere is $4\pi r^2$. So, the change in area with respect to time is . . ."

$$A' = \frac{dA}{dt} = 8\pi r \frac{dr}{dt}.$$

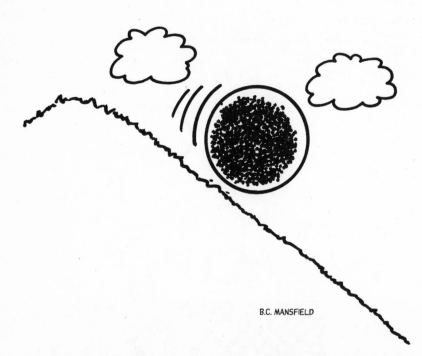

B.C. MANSFIELD

FIGURE 28. A meatball grows in surface area as it picks up debris while rolling down the hill. How fast is the meatball's *surface area* growing at the moment when the radius of the meatball is 10 cm?

"Next consider that the volume of a sphere is $V = (4/3)\pi r^3$. The change in volume with respect to time is . . ."

$$V' = \frac{dV}{dt} = 4\pi r^2 \frac{dr}{dt}.$$

I look at Fiona, her eyes wide with excitement. "Fiona, what do we know about the meatball?"

"We know the rate of volume increase for the meatball. You said the meatball increased 1 cubic centimeter per hour, so $dV/dt = 1$. That means $1 = 4\pi r^2(dr/dt)$."

"Correct. Using a little algebra, we can manipulate the equation so that it contains a term that is the derivative of the area. Watch this magic":

$$1 = 4\pi r^2(dr/dt) = (\tfrac{1}{2}r)[8\pi r(dr/dt)] = \tfrac{1}{2}r(A').$$

"When $r = 10$, $1 = \tfrac{1}{2}10(A')$. Therefore, the rate of change in surface area at the precise instant that the meatball is 10 cm in radius is $A' = 0.2$ cm^2 per hour. This is the value I was looking for—the change in the meatball's surface area." I pause. "At this rate, I wonder how long it would take for the meatball to be as large as the moon."

The three of us are silent for a while. I look at Fiona. "Fiona, would you like to have dinner together? I could get reservations for two at the Four Seasons, or perhaps you'd prefer something more exotic, like the new sushi place, the Green Wasabe, down on Broadway?"

Fiona pauses for a few seconds while considering my question. "Luigi," Fiona whispers, "that sounds nice, but should we exclude Rosario?"

Rosario puts up his hands. "Oh, go ahead. Enjoy yourselves. I want to rest because I know tomorrow you will be talking to us about integration. I just love the calculus integral symbol." He draws

$$\int.$$

Fiona gasps for breath as she gazes at the lovely integral symbol, so perfect, pristine, and austere. Oh God, I think. Someday I must build a home in the shape of an integral symbol.

"Just wait until tomorrow," I say. "As my mother used to tell me, the job of the integral is to transport us to new seas, while deepening the waters and lengthening horizons."

EXERCISES.........................

1. Luigi wants to be the first person to launch a meatball into outer space. He places an engine on the meatball so that it is flying at a speed of 200 kilometers per hour along a line making a 45-degree angle with the ground (Figure 29). How fast is the altitude *a* of the meatball changing?

2. How would your answer be different in problem 1 if the meatball were traveling at a 60-degree angle with respect to the ground?

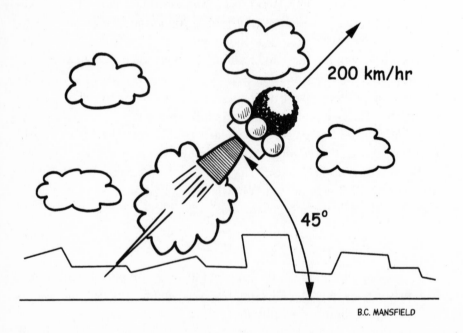

200 km/hr

45°

B.C. MANSFIELD

FIGURE 29. Launching a meatball into outer space.

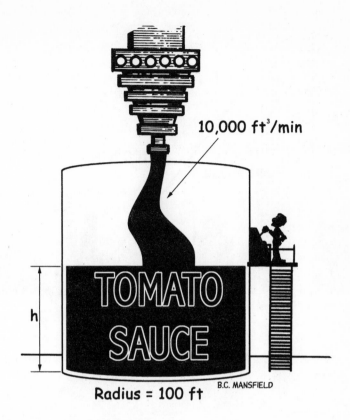

10,000 ft³/min

TOMATO
SAUCE

h

Radius = 100 ft

B.C. MANSFIELD

FIGURE 30. A huge can of tomato sauce.

3. In order to impress Fiona, Luigi has constructed a huge cylindrical can with a radius of 100 feet (Figure 30). The can is being filled with tomato sauce at the rate of 10,000 cubic feet per minute. How fast is the depth of the tomato sauce increasing? (Hint: The height of the sauce is h, and the volume of a cylinder is $\pi r^2 H$, where r is the radius and H is the height of the can.)

4. In problem 3, how fast would the depth of the tomato sauce rise if the can had a radius of 20 feet instead of 100?

5. Luigi has recently proposed the following experiment to his customers as a stimulus for the mind. Imagine a 100-mile long piece of train track leaning against a brick wall of a tall building as shown in Figure 31. The

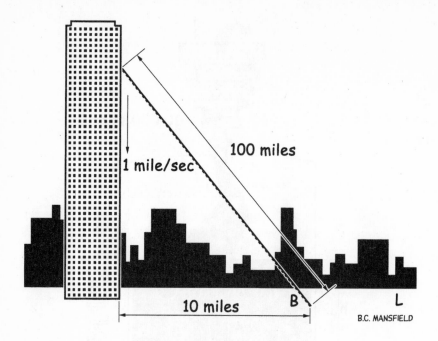

FIGURE 31. A very long train track leans against a tall building. Here *x* is 10 miles.

track is sliding down the wall at a speed of 1 mile per second. How fast is the bottom of the track, labeled *B*, moving toward Luigi, labeled *L*, when *B* is 10 miles away from the wall?

6. Refer to problem 5, but the track is 10 miles long and we want to know the speed of the bottom of the track when *B* is 5 miles away from the wall.

7. Luigi has invented a special kind of pizza dough that dramatically expands through time. First, he makes a tiny pizza only 1 inch in radius. He puts the minipizza on the counter and tells Fiona, "Watch this!" The radius of the pizza increases at the rate of 3 inches per minute. Fiona gazes at the growing pizza as the customers in the restaurant begin to scream in fear. Big Tony curses and runs from the restaurant before he is crushed by the turgid mass. How fast is the area of the pizza growing when the radius is 200 inches?

8. In problem 7, how fast is the pizza's area growing when the radius is 10 inches if the radius grows at half an inch per minute.

9. Luigi is watching an ant traveling along a countertop in his restaurant. The ant is wandering along the curve $y = x^2 + 3x - 1$. Luigi wants to dispose of the ant, but before he does, he is curious to learn at which points on the ant's path are the x- and y-coordinates of the ant changing at the same rate. Can you help Luigi?

10. Solve problem 9 for an ant traveling along the curve $y = x^2 + 4x - 3$.

11

Integration

> The great trick of regarding small departures from the truth as the truth itself—on which is founded the entire integral calculus—is also the basis of our witty speculations, where the whole thing would often collapse if we considered the departures with philosophical rigor.
>
> —Georg Christoph Lichtenberg, *Aphorisms*

A meatball is rolling down Fifth Avenue in New York City. It will roll for 1 hour. The police have blocked off the side streets to allow my experiment. People fill the sidewalks staring at this special event.

As the meatball passes Trump Tower, it gradually picks up speed. Donald Trump comes out of the tower and determines with special instruments that the velocity of the meatball as a function of time is

$$v(t) = 10t.$$

The velocity is in units of feet per hour.

I turn to Fiona. "What is $f(t)$?"

"What do you mean by $f(t)$?"

"What is the position of the meatball as a function of time? Because the velocity is the rate of change of position with respect to time, we can write $v(t) = f'(t)$."

"True," Fiona says.

"This means," I say, "that f is an *antiderivative* of v."

Fiona watches the cheering crowds as the meatball progresses. "What a strange word—antiderivative."

"Yes. Recall that $v(t) = 10t$. Can you think of any function $f(t)$ for which $10t$ is the derivative?"

"How about $f(t) = 5t^2$? In this case, we know that $f'(t)$ is $10t$."

"Fiona, correct! Actually, you could add any constant to $f(t)$ because we know that the derivative of a constant is zero and disappears from the expression."

Big Tony is with us, and he has brought a reclining chair to rest his body while he considers the traveling meatball. His chair has comfortable ergonomic padding and flip-up cup-holder armrests, one of which currently contains a large-size chocolate and banana milkshake.

Big Tony scratches his head. "So, $f(t)$ might equal $5t^2 + C$, where C is a constant?"

"Yes, $5t^2 + C$ is a perfectly good general solution. And at time zero, just as the meatball takes off at Central Park, C is the position of the meatball. Perhaps C is measured as the number of feet from one of the tips of Manhattan. From this point on, we know the position $f(t)$ of the meatball. Here are some antiderivatives of important functions." I write the adjacent table on Fifth Avenue.

Fiona stares at the table. "Let me try to verify one of these. I'll try to take the derivative of . . ." She writes on the pavement

$$\left(\frac{x^{r+1}}{r+1} + C \right)'.$$

ANTIDERIVATIVES

Function	Antiderivative
0	C
a	$ax + C$
x	$\frac{1}{2}x^2 + C$
x^r	$\frac{x^{r+1}}{r+1} + C \ (r \neq -1)$
x^{-1}	$\ln |x| + C \ (x \neq 0)$
$\cos ax$	$\frac{1}{a} \sin ax + C$
$\sin ax$	$-\frac{1}{a} \cos ax + C$
e^{ax}	$\frac{1}{a} e^{ax} + C$
$\frac{1}{\sqrt{a^2 - x^2}}$	$\arcsin \frac{x}{a} + C$
$\frac{1}{a^2 + x^2}$	$\frac{1}{a} \arctan \frac{x}{a} + C$

She takes the derivative of this expression, and sure enough, the derivative is x^r.

Big Tony seems to be dozing off. I tap him on the shoulder. "Big Tony, here is your next problem."

"Can't a fellow get any sleep around here?"

"Find the antiderivative of the following":

$$3x^2 - 7x.$$

Big Tony considers the problem. "Using the fourth rule in your table of rules, the antiderivative is . . ."

$$3\frac{x^3}{3} - 7\frac{x^2}{2} + C = x^3 - \frac{7}{2}x^2 + C.$$

"Good," I say. "Let's go for a walk." We leave Big Tony's recliner behind for Rosario to pack in the truck, as Fiona, Big Tony, and I explore Fifth Avenue. After half an hour, the city's fashionable shopping section has given way to a series of smaller shops. I walk slowly, passing windows of secondhand clothing stores trying to establish themselves as grunge boutiques. Some of the store signs read CALCULUS IS COOL in honor of Calculus Day, which I convinced the city to hold once a year (Figure 32). One of the stores is called Mathematicians Unite, and it sells a startling array of goods—calculus playing cards, nightgowns embroidered with mathematical formulas, calculus bubblegum, and bronze busts of Newton and Leibniz displayed on black velvet. I stop for a few minutes in front of the store.

I turn to Fiona. "Most of our previous lessons focused on differentiation, but for the remainder of our discussions we'll focus on the other part of calculus: integration. Integration is in many ways the reverse of differentiation."

Fiona and Big Tony sit down on a bench. Sometimes they watch the energetic customers inside Mathematicians Unite, but most of the time they pay careful attention to my discussions.

"Let me tell you about integrals. There are two types of integrals: the *definite integral* and the *indefinite integral*. The antiderivatives on which we've just been working are examples of indefinite integrals of a function. The process is the opposite of taking a derivative. Indefinite integrals of a function yield another function. Definite integrals of a function produce a specific number. This will become clear soon."

A woman in the store, perhaps the manager, has a gaudy punk hairdo— half purple, half maroon—and looks as if she might weigh around ninety

B.C. MANSFIELD

FIGURE 32. Calculus is cool and so is Luigi.

pounds. She looks at the three of us and grins, revealing two large white canine teeth. They remind me of the teeth of a vampire.

I turn away from the woman. "Integrals are amazing. For example, integrals are good for finding areas under curves and also for finding volumes. Let's consider a parabola, $y = x^2$. We want to find the area under the graph of this curve in the interval between $x = -1$ and $x = 1$. One way to estimate this would be to draw a set of thin rectangles beneath the curve and add the areas."

I draw a figure of a parabola-like curve on the sidewalk using a piece of chalk (Figure 33). As I draw, a tall man in a ragged coat and long, scraggly beard comes by and listens. From across the street, I also see a policeman staring at me with a questioning look.

"The height of each rectangle is $f(x)$. The sum of the area of the rectangles is called the Riemann sum, and it can be represented mathematically as . . ." I draw on the sidewalk

$$\sum f(x)\Delta x.$$

The tall man with the beard nods.

I point to the equation. "The term Δx corresponds to the horizontal thickness of each rectangle. Therefore, the term $f(x)\Delta x$ corresponds to the area of

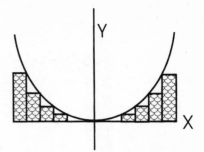

FIGURE 33. Approximating the area under a parabola-like curve using rectangular areas.

a rectangle. The Σ symbol tells us we want to sum the areas of *all* the rectangles. The thinner the rectangles become, the more accurately we can estimate the area under the curve. We can compute the exact area in the limit as the width of each rectangle, Δx, approaches zero. Mathematically, this looks like . . ."

$$\text{Area under curve} = \lim_{\Delta x \to 0} \sum f(x)\Delta x.$$

"This limit is so special that it is given a name, the *Riemann integral*, or definite integral. It also has a special symbol that looks like a stretched-out S":

$$\int_a^b f(x)dx = \lim_{\Delta x \to 0} \sum f(x)\Delta x.$$

"The a and b represent the interval over which we are calculating the sum. In our example with the parabola, a is –1 and b is 1. In this mathematical representation, note how Δx is written as dx."

The policeman who has been watching us steps closer. "Using the Riemann integral," he says, "we can obtain the area under $y = f(x)$ between points $x = a$ and $x = b$."

"That's right, officer," I say. "The *fundamental theorem of calculus* tells us how we can compute $\int_a^b f(x)dx$ for any continuous function f between $x = a$ and $x = b$. All we have to do is find any antiderivative F of f. Once we know the antiderivative, then we know . . ."

$$\int_a^b f(x)dx = F(b) - F(a) = \left[F(x)\right]_a^b.$$

I look from the officer to Fiona. "The notation $\left[F(x)\right]_a^b$ is just a more compact way of writing $F(b) - F(a)$."

"This is a bit confusing," Fiona says. "Can we see an example?" She brushes back her hair that the faint wind has caressed out of place.

"Sure. Let's try a simple example. Again, we want to find the area under $y = x^2$ for $-1 \leq x \leq 1$. Here $f(x) = x^2$, and the typical antiderivative is $F(x) = (1/3)x^3 + C$. The area is therefore . . ."

$$\int_a^b x^2 dx = \frac{1}{3}x^3 + C\rfloor_{-1}^{+1} = \left(\frac{1}{3} \times 1^3 + C\right) - \left(\frac{1}{3} \times (-1^3) + C\right) = \frac{2}{3}.$$

The bearded man applauds, evidently impressed by the barrage of impressive-looking symbols. I nod at him.

"The constants C just cancel, so we don't usually have to write them down in such problems." I am momentarily distracted by the bearded man who spits a wad of chewing tobacco on the sidewalk. I don't like the way some of his spit has defaced the integral sign, but perhaps it was an unintentional affront to my dignity.

"Gross," says Fiona.

Big Tony starts to flex his massive biceps in an apparent attempt to intimidate the tall man. "*Ben fattom*," Big Tony whispers under his breath. "*Ottimo lavoro.*"

"Pardon me," says the man. "I didn't mean to spit so close to the equation."

I cough to get everyone's attention. "It is customary to represent the antiderivative of $f(x)$ by an indefinite integral $\int f(x)dx$—that is, an integral sign without limits. The Riemann integral $\int_a^b f(x)dx$ is called the definite integral."

I look at Fiona. "Let's try another example with the definite integral. Fiona, solve this one":

$$\int_a^b (3x^2 - 4x)dx.$$

"Okay," she says. "First we calculate the antiderivative using the rules you taught us." She writes on the sidewalk

$$\int (3x^2 - 4x)dx = \frac{3x^3}{3} - \frac{4x^2}{2} + C = x^3 - 2x^2 + C.$$

The policeman nods.

Fiona continues. "Now we evaluate this between $x = 1$ and $x = 3$":

$$\int_1^3 (3x^2 - 4x)dx = x^3 - 2x^2 \big]_1^3 = (27 - 18) - (1 - 2) = 10.$$

"That's right!" I say. "Notice how we don't have to worry about adding the arbitrary constant C when we solve these kinds of problems." I pause. "Here are some of the most basic rules for definite integrals." I take a card from my pocket. On the card is a table listing several rules for integration. Fiona, Big Tony, the tall man, and the policeman pass the card around. Is that a glimmer of awe I see in their eyes?

RULES FOR INTEGRATION

Rule	Notation
Sign switch	$\displaystyle\int_a^b f(x)dx = -\int_b^a f(x)dx$
Sum	$\displaystyle\int_a^b \big[f(x) + g(x)\big]dx = \int_a^b f(x)dx + \int_a^b g(x)dx$
Constant multiplier	$\displaystyle\int_a^b cf(x)dx = c\int_a^b f(x)dx$
Intervals	$\displaystyle\int_a^c f(x)dx = \int_a^b f(x)dx + \int_b^c f(x)dx$
Area between two curves	$\displaystyle\int_a^b \big[f(x) - g(x)\big]dx$

"Fiona, do you recall the table of antiderivatives I showed you a few minutes ago?"

"It seems like a lifetime ago. But how could I ever forget?"

I nod. "You have to be careful when you use the formulas. If the terms in the integrand are more complicated, for example, if the xs are replaced by more complicated functions, you'll have to remember to use the chain rule."

"I don't get what you're saying," Fiona says.

"Let me show you what I mean. If I ask you what the integral of x^3 is, you would tell me . . ." I write on the sidewalk

$$\int x^3 dx = \frac{x^4}{4} + C.$$

"You can confirm this by . . ."

$$\left(\frac{x^4}{4}\right)' = x^3.$$

"We simply used the rule in the table that says the antiderivative of x^r is . . ."

$$\frac{x^{r+1}}{r+1} + C$$

or

$$\int x^r dx = \frac{x^{r+1}}{r+1} + C.$$

"But what if I asked you to find the integral of $(x^3+1)^2$? We can't simply plug this into our formula. Why? Because here I have replaced x in our simple expression with something more complicated, namely $x^3 + 1$. In these cases, you would be wise to study the following table." I pass around another card.

ANTIDERIVATIVES

Indefinite Integral	Antiderivative
$\int u^r du$	$\dfrac{u^{r+1}}{r+1} + C \ (r \neq 1)$
$\int u^{-1} du$	$\ln\lvert u\rvert + C \ (u \neq 0)$
$\int \cos u\,du$	$\sin u + C$
$\int \sin u\,du$	$-\cos u + C$
$\int e^u du$	$e^u + C$
$\int \dfrac{du}{\sqrt{a^2 - u^2}}$	$\arcsin \dfrac{u}{a} + C$
$\int \dfrac{du}{a^2 + u^2}$	$\dfrac{1}{a} \arctan \dfrac{u}{a} + C$

"Fiona, in order to solve more complicated integrals with functions instead of single variables, we must use the analogous rule":

$$\int u^r du = \frac{u^{r+1}}{r+1} + C.$$

"Here, as in all the integral rules for more complicated functions, $du = (du/dx)dx$, which means you need the du in the integrand to make the integration straightforward. For example, find the integral of the following":

$$\int (x^3 + 1)^2 3x^2 dx.$$

Fiona says, "Here the trick seems to be that I have to recognize that $u = x^3 + 1$, $r = 2$, and $du = 3x^2 dx$." She draws brackets around u and du to help everyone see them more clearly:

$$\int \left[(x^3 + 1) \right]^2 \times [3x^2 dx].$$
$$\quad\;\; u \qquad\qquad du$$

Fiona continues. "Using the previous formula, we find . . ."

$$\int (x^3 + 1)^2 3x^2 = \frac{(x^3 + 1)^3}{3} + C.$$

The policeman applauds.

I smile. "Yes, you needed the 'extra' $3x^2$ term to use the integral formulas. For example, you cannot directly use the formulas to solve the following":

$$\int (x^3 + 1)^2 dx.$$

"Many calculus procedures involve pattern recognition—you have to *see* a possible *du* so that you can easily visualize how to fit equations to the rules we know. Sometimes you'll even need to refer to large tables of integrals that have dozens of patterns and tricks, and in these instances you also need to see how to fit your formula into existing structures. And some of the integrals just can't be solved using any tables. I'm only giving you the basic rules here."

The tall, disheveled man says, "Darn, I had hoped to see more integral rules today. I think Fiona would appreciate seeing them."

I look at the man. "Who the heck are you?"

The man pays no attention to me. I try to ignore him. He gives me the creeps. The calculus tattoos on his arms are not unattractive, except for the fact that they are accompanied by skulls and body piercings.

I turn to Big Tony. "Here's another to try":

$$\int 3e^{x^3} x^2 dx = ?$$

Before Big Tony has a chance to consider my question, the man with the tattoos grabs the chalk from Big Tony and says, "This is trickier. But, if we try really hard, we recognize that we can substitute $u = x^3$, and $du = 3x^2 dx$, or $(1/3)du = x^2 dx$. Now let's make the substitutions . . ."

$$\int 3e^{x^3} x^2 dx = 3 \int e^u \frac{1}{3} du = \int e^u du = e^u + C = e^{x^3} + C.$$

I grab back the chalk from the man with the long, scruffy beard. "You are right," I say grudgingly.

I clap my hands together. "People, we're almost done with today's lesson. Many integrals look similar to the ones we can antidifferentiate if we adjust them by substituting another variable for x, as we did here. Sometimes if you see $ax + b$, you might substitute $u = ax + b$. Or if you see x^p, you might substitute $u = x^p$. There are lots of tricks to try. As a final example, consider a construct like this":

$$\int \frac{1}{4 + x^2}\, dx.$$

"We can use the last rule in the antiderivative table to solve this. Recall the rule? It said . . ."

Function	Derivative
$\dfrac{1}{a^2 + x^2}$	$\dfrac{1}{a}\arctan\dfrac{x}{a} + C.$

"So, here a is 2":

$$\int \frac{dx}{4 + x^2} = \frac{1}{2}\arctan\frac{x}{2} + C.$$

Big Tony is yawning. "It's late."

I look at my watch. "You're right. It's time to get back to the restaurant and go to bed. You've learned a lot today." To save money on New York's very high rents, I have temporarily given Big Tony and Fiona separate rooms above the restaurant.

As we make our way back up Fifth Avenue, much of the crowd that had gathered to watch the rolling meatball has dispersed. I wonder about its ultimate fate.

Once back at the restaurant, Fiona and I follow Big Tony up the stairs to his room, as we hold handkerchiefs against our mouths and noses. The smell of Big Tony's place is much worse than I remember it. It is a heavy smell, a smell of putrefaction. Probably the smell of some of his leftovers, or Limburger cheese, or his latest cooking.

After saying good-night to Big Tony and then to Fiona, I go to my room. Fiona's minty perfume definitely leaves a lasting impression, and it covers up the overpowering odor of Big Tony's garlic and aged cheese.

A pizza chef sometimes feels very alone.

I stare out the window. Darkness comes snaking across the top floor of the restaurant. I put on headphones and turn on a CD. It is *Nessun Dorma* by Luciano Pavarotti and the John Alldis Choir. I am sleepy. The music pours into the room like a good wine—like a streaming river, water I have looked into, water I have held.

EXERCISES........................

Antiderivatives (Indefinite Integrals)

Find the following antiderivatives:

1. $\int \left(\pi + \dfrac{2}{\sqrt{x}} \right) dx$

2. $\int \left(\pi + \dfrac{2}{\sqrt{x}} \right) dx$

3. $\int (2\sin x - 4\cos x + 2x^3)\,dx$

4. $\int (6\cos x - 4\sin x - 3x^4)\,dx$

5. $\int \sqrt{5x - 2}\,dx$

6. $\int \sqrt{2x + 4}\,dx$

Definite Integrals

Find the following definite integrals:

7. $\int_{1}^{3} (6x^2 + x + 1)\, dx$

8. $\int_{1}^{3} (9x^2 + 2x - 1)\, dx$

9. $\int_{0}^{1} \dfrac{x}{(3x^2 - 1)^3}\, dx$

10. $\int_{3}^{4} \dfrac{x}{(2x^2 + 4)^3}\, dx$

12

Logarithmic Differentiation, Integration by Parts, Trigonometric Substitution, and Partial Fractions

Was Calculus invented or discovered
—Was it made or was it uncovered?
Some only deride,
Others try to decide.
Still we aren't sure if it's one or the other.

—Jan Gullberg, *Mathematics: From the Birth of Numbers*

Today Fiona's earrings are in the shape of integral symbols. Her shirt has a picture of Gottfried Wilhelm von Leibniz, one of the discoverers of calculus. Underneath his smiling face are printed the words I LOVE CALCULUS.

Fiona, Big Tony, and I are seated in the back of my restaurant. I turn to Big Tony. "Sometimes when we have a very tough function to differentiate, we take the natural log of both sides of the equation before taking the derivative. This makes our lives a lot easier."

"Why?" Big Tony asks as he bites into a meatball and garlic calzone the size of a football.

"Let's assume you want to take the derivative of the following . . ." I write on the wall

$$F(x) = (x^6 + 6)(x^7 + 7)(x^8 + 8)(x^9 + 9).$$

Big Tony takes a look. "It would be a real pain in the butt to apply the product rule many times to this."

"Correct. If you want to save yourself a whole lot of pain, first take the log of both sides":

$$\ln f(x) = \ln[(x^6 + 6)(x^7 + 7)(x^8 + 8)(x^9 + 9)].$$

"Because we know that $\ln(ab) = \ln(a) + \ln(b)$, our big product becomes a sum":

$$\ln f(x) = \ln(x^6 + 6) + \ln(x^7 + 7) + \ln(x^8 + 8) + \ln(x^9 + 9).$$

From the side of the restaurant, a woman in a sleek black dress applauds upon seeing this equation. She waves her vegetarian pizza slice in our direction. I recognize the woman. Her name is Carmela. I smile at her. Fiona gives her a dirty look.

"What do we do next?" Fiona asks.

"Now we differentiate both sides. Recall that I once gave you a list of rules, including $(\ln u)' = u'/u$. This means we can rewrite our particular equation as . . ."

$$\frac{f'(x)}{f(x)} = \frac{6x^5}{x^6+6} + \frac{7x^6}{x^7+7} + \frac{8x^7}{x^8+8} + \frac{9x^8}{x^9+9}.$$

"Our mission is to find $f'(x)$, so we must multiply both sides of the equation by $f(x)$":

$$f'(x) = f(x)\left(\frac{6x^5}{x^6+6} + \frac{7x^6}{x^7+7} + \frac{8x^7}{x^8+8} + \frac{9x^8}{x^9+9}\right).$$

or

$$f'(x) = (x^6 + 6)(x^7 + 7)(x^8 + 8)(x^9 + 9)\left(\frac{6x^5}{x^6 + 6} + \frac{7x^6}{x^7 + 7} + \frac{8x^7}{x^8 + 8} + \frac{9x^8}{x^9 + 9}\right).$$

"And this is our answer!" Carmela cries out. "You've found the derivative."

"Wasn't that fast?" I ask as Fiona uses her hands to turn my head away from Carmela and back to the equation on the wall. "Just imagine," I say, "the mess we would have trying to do this without logarithmic differentiation or by trying to use the product rule or multiplying all the terms."

I watch Big Tony as he devours his calzone like there is no tomorrow. He is already half finished. I turn to Fiona. "This approach is called *logarithmic differentiation*. Let's use logarithmic differentiation to find the derivative of something that looks simple, such as $f(x) = 3^x$."

"Let me give it a try," Fiona says. "First we take the logs of both sides":

$$\ln[f(x)] = \ln(3^x).$$

"By the law of logarithms, which I once learned in my algebra class, we get . . ."

$$\ln(f(x)) = x \ln 3.$$

I nod, and Fiona continues.

"Notice that ln 3 is a constant, so taking the derivative of each side gives . . ."

$$\frac{f'(x)}{f(x)} = \ln 3.$$

"That means . . ."

$$f'(x) = 3^x \ln 3.$$

Fiona raises her marker high into the air, as if she is a triumphant warrior returning from a glorious victory.

I clap my hands together. "You did it!"

Fiona looks in Carmela's direction. Carmela does not applaud.

Big Tony is uncharacteristically silent. Perhaps the calzone is too much for him. He is three-fourths finished. At last he speaks.

"Luigi, what do we do if we want, in general, to take the integral of a product of two functions."

"Unfortunately, the integral of a product is not the product of the integrals."

Fiona nods. "I remember from an earlier lesson that the derivative of a product is also not just the product of the derivatives. We had to use the product rule for this."

"Correct. The product rule for integration—also called *integration by parts*—looks like this." I write on the wall

$$\int u\,dv = uv - \int v\,du.$$

"The challenge for using this method is always the challenge of assigning u and v to functions in the original integral."

"Practice makes perfect," Fiona says. "Let's solve a few examples."

"Okay, find the antiderivative for this":

$$\int x \sin x\,dx.$$

"Let's see what happens," I say, "if we let $u = x$ and $dv = \sin x\,dx$. This means that $du = dx$ and $v = -\cos x$. Using the rule for integration by parts (which is a product rule for integration), we get . . ."

$$\int x \sin x\,dx = x \cos x + \int \cos x\,dx = -x \cos x + \sin x + C.$$

"Great," says Fiona.

I nod. "Let's try another":

$$\int \frac{\ln x}{x^2}\,dx.$$

"Ooh, that looks complicated," she says.

Carmela stands up. "Not too complicated," she says. Carmela holds out her hand to Fiona, motioning for the marker. After ten seconds Fiona slowly gives the marker to Carmela.

Carmela starts writing on the wall. "Let $u = \ln x$ and $dv = (1/x^2)dx$. Then, $du = (1/x)dx$ and $v = -1/x$."

"How did you get that for v?" Big Tony asks. He is seven-eighths finished with his calzone.

"Consider $dv = x^{-2}dx$," Carmela says. "The antiderivative of dv is $-x^{-1}$."

Fiona pries the marker from Carmela's hand and begins to write on the board as she says, "Let me take it from here. Using Carmela's values for u, du, v, and dv, we have . . ."

$$\int \frac{\ln x}{x^2}\,dx = \int \overset{u}{(\ln x)} \times \overset{dv}{\left(\frac{1}{x^2}\,dx\right)}$$

$$= \ln x \times \left(-\frac{1}{x}\right) - \int \left(-\frac{1}{x}\right) \times \left(\frac{1}{x}\,dx\right) = \frac{-\ln x}{x} + \int \frac{1}{x^2}\,dx$$

$$= \frac{-\ln x}{x} - \frac{1}{x} + C = -\frac{1}{x}(\ln x + 1) + C.$$

Fiona looks at the solution and says, "Yes, I am good."

Big Tony says, "Way to go!"

Then Big Tony turns to me. "Luigi, it seems like you have been guessing at what part of the integrand should be u and what part should be dv. These appear to be really hard guesses to make."

"It's not exactly guessing. Notice how the dv part has to be something you are able to integrate. The remaining part of the expression is u. Also, look at the tail end of the integration by parts rule." I write

$$\int v\,du$$

"I see it," Big Tony says.

"If this tail end isn't easier to integrate than the original expression you're trying to integrate, think about trying other assignments for u and dv."

Big Tony is chewing on the last bit of his calzone. I expect he will finish just as our lesson concludes. He always seems to time his eating perfectly.

I turn to Fiona. "Now, I want to tell you about another integration method. It's called *trigonometric substitution*." I take a card out of my pocket and hand it to Fiona (Figure 34).

"Impressive card," Carmela says. "May I have one too?" She flashes her white teeth, and I notice she has a large diamond ring on her finger. Perhaps her ex-husband gave it to her.

Fiona snaps her fingers. "That's his last card." Then she turns to me. "Luigi, do you give that card to everyone you teach?"

"Ah, years ago I regularly handed the card out. But I haven't given it to anyone else but you in the last two years." We stare at each other for several seconds.

"Fiona," I say, "here is an example of how you would use the trigonometric method. Consider this . . ."

B.C. MANSFIELD

FIGURE 34. Luigi, a man prepared for all occasions.

$$\int \frac{1}{\sqrt{1-x^2}}\,dx.$$

Big Tony says, "Yesterday, you gave us a rule for this":

$$\int \frac{du}{\sqrt{a^2-u^2}} = \arcsin\frac{u}{a} + C.$$

I nod. "But I want you to solve it without the rule today in order to make a point."

"This looks painful to solve," Fiona says.

"Yes, it looks that way, especially if you don't know all the antiderivative rules I gave you earlier. Let's find an easy way to solve it using trigonometric substitution. Recall the following from your basic trigonometry":

$$\sin^2\theta + \cos^2\theta = 1.$$

"If we let $x = \sin\theta$, we can gradually eliminate the difficult denominator. Let's start . . ."

$$\int \frac{1}{\sqrt{1 - \sin^2\theta}} \, dx.$$

"From our trigonometric rule, we know that . . ."

$$\sin^2\theta = 1 - \cos^2\theta.$$

"So, the denominator becomes $\sqrt{\cos^2\theta}$ or simply $\cos\theta$. So we now have . . ."

$$\int \frac{1}{\cos\theta} \, dx.$$

Carmela touches the equation. "Luigi, but now we have a dx and a θ in our expression. That looks odd."

"Yes, that's a mismatch for integration, and we would like the variables to be the same. One way to express everything in terms of thetas is to recall our initial substitution $x = \sin\theta$. If we take the derivative of both sides of $x = \sin\theta$ with respect to $d\theta$, we get . . ."

$$\frac{dx}{d\theta} = \cos\theta$$

or

$$dx = \cos\theta d\theta.$$

"The integral finally becomes . . ."

$$\int \frac{1}{\sqrt{1 - x^2}} \, dx = \int \frac{1}{\cos\theta} \cos\theta \, d\theta = \int 1 d\theta = \theta + C.$$

"Hooray!" Fiona yells.

"Wait, we are not done. We want to express our answers with xs, not thetas. We know that $x = \sin\theta$, which means arcsin $x = \theta$. And we have our solution":

$$\int \frac{1}{\sqrt{1 - x^2}} \, dx = \theta + C = \arcsin x + C.$$

"And this," I say, "is an example of solving an integral by trigonometric substitution."

Fiona, Big Tony, Carmela, and I are silent for a minute. A breeze blows through the open window carrying with it the aroma of roses and peppermint from the nearby garden.

"Fiona," I say, "do you like my new table?" I motion for her to look down at the special heat-sealed table edge and high-pressure laminate that provides a seamless edge. This means no food entrapment or bacterial buildup.

"Nice," Fiona says as she trails her fingers along the table's ultrasmooth contours. "It appears to be stain resistant, and the colors won't degrade after lengthy exposure to the sun."

Big Tony looks up and says, "Superior impact and puncture resistance. Semi-soft edge material saves wear and tear on the table edges and legs. Rich textured finish. Also flame retardant."

Carmela says, "The finish enhances the natural grain and fine patterns of the wood."

A shiver goes up my spine. "Why are we all talking in such a stilted, unnatural way?"

"It's as if—" Fiona pauses.

"What?" I ask.

"As if we are characters in a book and the author is a little nuts."

I chuckle at that one. "Fiona, you do have a great imagination."

I notice that Big Tony has a single bite left of his calzone. I turn to Fiona. "Now we will discuss a final integration method: *partial fractions*." These words delight Fiona.

"Take a look at this tough problem," I say:

$$\int \frac{2x+1}{x^2+4x+3} \, dx.$$

"Ordinarily, we'd have a heck of a time trying to integrate this. However, with the method of partial fractions, the integral is really not too bad. To start, we are going to break this integral into two separate integrals. The first step is to represent the complicated fraction as two simpler fractions. For example, we can factor the denominator as $x^2 + 4x + 3 = (x + 1)(x + 3)$. We can then express the fraction in the integral as . . ."

$$\frac{2x+1}{(x+1)(x+3)}.$$

"Next, we express this as two separate fractions with numerators A and B. We must solve for A and B":

$$\frac{2x+1}{(x+1)(x+3)} = \frac{A}{(x+1)} + \frac{B}{(x+3)}.$$

"We can get rid of the denominators simply by multiplying all terms by $(x + 1)(x + 3)$, so . . ."

$$2x + 1 = A(x + 3) + B(x + 1).$$

"Let me give this a try," says Fiona. "Let's find some values for A and B. First we clear out the parentheses":

$$2x + 1 = Ax + 3A + Bx + B.$$

Fiona looks at me. I nod, and she continues. "Gathering terms gives . . ."

$$2x + 1 - Ax - 3A - Bx - B = 0$$

$$2x - Ax - Bx - B - 3A + 1 = 0$$

$$(2 - A - B)x - B - 3A + 1 = 0.$$

"Because x can have any value, let's try $x = 0$. When $x = 0$, we have . . ."

$$-B - 3A + 1 = 0$$

or

$$\boxed{B + 3A = 1.}$$

Now Fiona looks at Carmela and raises her voice slightly. "This also implies that $(2 - A - B)x = 0$." Fiona pauses. "Again, let's choose a simple value for x. Here $x = 1$. This gives . . ."

$$2 - A - B = 0 \text{ or } -2 + A + B = 0$$

or

$$\boxed{A + B = 2.}$$

Carmela puts up her hand. "Fiona, wait. I want a chance." Carmela continues the discussion. "We have two sets of equations that must be true. See my previous equations in boxes." Her hand races across the wall as she scrawls equations with increasing abandon:

$$A + B = 2$$

$$B + 3A = 1$$

or

$$B = 1 - 3A.$$

Carmela is unstoppable. Her breathing becomes quicker. "Thus," she says . . .

$$A + (1 - 3A) = 2 \text{ or } -2A = 1 \text{ or } A = -\frac{1}{2},$$

"which means that $B = 5/2$."

"Okay," I say. "Very good. Let me finish up." I remove the marker from Carmela's trembling hand. "Now that we have values for A and B, we can finally write the initial integral as . . ."

$$\int \frac{2x+1}{(x+1)(x+3)} dx = \int \frac{-1/2}{(x+1)} dx + \int \frac{5/2}{(x+3)} dx.$$

"Recall that . . ."

$$\int \frac{1}{x} dx = \ln|x| + C.$$

"So we get as our final answer . . ."

$$-\tfrac{1}{2}\ln |x + 1| + (5/2)\ln |x + 3| + C.$$

Tony has finished his calzone. All is right with the world.

■ ■ ■

A few minutes later, Fiona is sipping a huge goblet of iced tea in which a sprig of crushed mint floats. Carmela and I decide to have the same beverage except we have tarragon powder added to ours.

I turn to Big Tony. "What are you waiting for?"

"They are preparing my latest drink."

"Uh oh. I don't like the sound of it."

"*Titan arum* tea—perhaps you've heard of it?"

"*Titan arum!*" Fiona says. "That's the smelliest flower in the world."

I nod. "In fact, it is one of the strangest flowers on earth. It mimics the texture and smell of decaying flesh to attract the carrion beetles and flies that it needs for pollination. It's also known as *Bunga bangkai* or the corpse flower. You can smell it from eighty feet away. But boy does it taste good. Sweeter than honey."

Fiona gets up from the table and rolls her eyes.

Big Tony studies our expressions for a few seconds, and says, "Sir, just kidding! I don't drink *Titan arum* tea."

"Kidding?" Carmela says as she touches Big Tony's arm. "I thought you never kid."

"I'm trying to learn," he says.

EXERCISES......................

1. Use logarithmic differentiation to find the derivative of

$$y = x^3 \sqrt{2 + x^2}.$$

2. Use logarithmic differentiation to find the derivative of

$$y = x^3 \sqrt{20 - 5x^2}.$$

3. Use integration by parts to find

$$\int x e^{4x} dx.$$

4. Use integration by parts to find

$$\int x e^{5x} dx.$$

5. Use the method of partial fractions to find

$$\int \frac{dx}{x^2 - 16}.$$

6. Use the method of partial fractions to find

$$\int \frac{dx}{x^2 - 25}.$$

7. Use logarithmic differentiation to calculate the derivative of $f(x) = (x^6 + 666)(x^7 + 777)x^8$.

8. Use logarithmic differentiation to calculate the derivative of $f(x) = (x^{20} - \pi)(x^7 - 777)x^3$.

9. Find the following integral using integration by parts:

$$\int \left(\frac{\ln x}{\pi^2 x^2} + \frac{2x^3}{\pi^2} \right) dx.$$

10. Find the following integral using integration by parts:

$$\int \left(\frac{e^3 \ln x}{x^2} + 25e^3 x^3 \right) dx.$$

Exponential Growth and Decay

At the outset of differentiation
We needed strong dedication.

—Jan Gullberg, *Mathematics: From the Birth of Numbers*

Fiona looks at the broken pizza oven. The electrical cord is completely severed and a dial is shattered. "Luigi, how will we fix it?"

"Fiona, as my mother always used to tell me, 'If it isn't broken, don't try to fix it.'"

Fiona looks at me with wide eyes. "But the pizza oven *is* broken. Your mother's aphorism does not apply in this situation."

My mind starts churning, trying to think of a resolution to Fiona's logical dilemma. I think of integrals, derivatives, and limits—but all to no avail. I decide the safest approach is to change the subject.

"Fiona, let's sit down and talk calculus." I motion for her to follow me to a nearby table that has a pizza resting on it. The pizza is covered by a metallic dish. "Many phenomena in our world exhibit a kind of rapid growth called *exponential growth*. For example, the bacteria on this pizza are doubling their population every twenty minutes."

Rosario and Big Tony sit down. Rosario says, "What pizza are you talking about?"

"This one," I say as I remove the metallic covering.

On the table near Rosario, Big Tony, and Fiona is a nearly liquid mass of horrid putrescence.

"You're a sick man!" screams Fiona.

Big Tony leaps from his chair. "Cover it up," he says.

"What is exponential growth?" Rosario asks me.

"Growth in which the rate of change of a function is proportional to the value of the function. Or the more bacteria, the faster the population grows. The opposite of exponential growth is *exponential decay.* Radioactive substances exhibit this kind of decay. The less of the radioactive substance present, the slower it decays" (Figure 35).

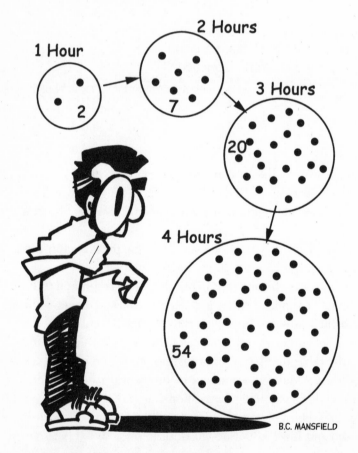

1 Hour

2 Hours

2

7

3 Hours

20

4 Hours

54

B.C. MANSFIELD

FIGURE 35. An example of exponential growth.

I cover up the pizza and begin to write on the wall. "We can describe this kind of growth or decay in the following equation":

$$\frac{dN}{dt} = kN.$$

"Assume that N is the number of bacteria on the pizza. The rate of *change* in the number of bacteria is proportional to the *number* of bacteria. The number k is just a proportionality constant. It's positive when there is growth and negative when the bacterial population decreases in number. Sometimes physicists write the derivative of N as $\dot{N}$, so the formula might also be written as $\dot{N} = kN$. These expressions are known as *simple differential equations* because the equations involve derivatives. Let's try to solve this equation in order to determine a value for N as a function of time. If we can do this, we can estimate the number of bacteria on this pizza as a function of time."

I look at Fiona, who is staring at the metal dish atop the pizza. "Here we can treat the derivative as if it is a fraction," I say, "and we can manipulate the equation to obtain . . ."

$$\frac{dN}{N} = kdt.$$

"Next, we integrate both sides":

$$\int \frac{1}{N} dN = \int kdt$$

$$\ln|N| = kt + C.$$

"We can raise both sides of the equation to the power of e to get $e^{[\ln|N|]} = e^{[kt + C]}$, and we can get rid of the absolute value signs because the number of bacteria is always positive":

$$N = e^{kt + C} = e^{kt}e^{C} = Ae^{kt}$$

where

$$A = e^{C}.$$

"Our exponential growth or decay equation is simply the following":

$$N(t) = N_0 e^{kt}.$$

"Here, N_0 is the initial number of whatever is decaying or growing."

"Where does N_0 come from?" Fiona asks.

"Let's study our previous equation, $N = Ae^{kt}$. What does this tell us happens at $t = 0$?"

"Well," Fiona says, "it tell us that $N(0) = Ae^{k0}$."

"Right. And because any number raised to the power of zero equals one, we have $N(0) = N_0 = A$. This gives us our N_0 number in the formula." I pause. "Rosario, would you be so kind to bring out a fresh pizza?"

"Certainly, sir." Rosario goes into the kitchen and returns with a pizza that is a few degrees above room temperature.

From my pocket, I withdraw a small amber vial. "Inside this vial are 10 fruiting bodies of the slime mold *Dictyostelium discoideum*. Under a magnifying glass, the fruiting bodies resemble little balls on top of a thin stalk, almost mushroomlike in appearance." I open the vial and place the 10 fruiting bodies in the center of the pizza.

"This is a genetically enhanced slime mold," I say. "It multiplies quickly. The genetic alterations, alas, also make the slime molds emit a foul stench, but we need not worry about that for now." I wait for a minute. "By tomorrow at this time there will be 2 million slime mold bodies on the pizza. The pizza can contain a total of 6 billion slime molds, and our goal is to determine when the pizza will contain this number of slime molds. First, let's start with our exponential growth equation":

$$N(t) = N_0 e^{kt}.$$

"*Now* is at time $t = 0$, and N_0 is 10. So we have . . ."

$$N(t) = 10e^{kt}.$$

"The next step is to determine k. We know that tomorrow at this time, which corresponds to 24 hours, the population will be 2 million slime mold bodies." I write on the wall

$$N(24) = 2,000,000 = 10e^{k(24)}$$

or

$$200,000 = e^{k(24)}.$$

"Let's solve for k":

$$\ln(200,000) = 24k$$

$$k = \ln(200,000)/24 \approx 0.508.$$

"Now that we have the value for k, we can substitute this value in our original equation":

$$N(t)=10e^{0.508t}.$$

Fiona asks, "Does this tell us the number of bacteria on the pizza at any time?"

"Yes."

Fiona stares at the pizza. "When will the pizza have reached its maximum of 6 billion slime molds?"

I don't immediately answer Fiona's question. "Fiona, as we chat, the slime molds have already started their rapid multiplication. Notice that people in the restaurant are beginning to gag from the stench, which resembles that produced by a rotting water buffalo corpse, or so I'm told. I have never actually encountered a rotting water buffalo corpse."

Fiona covers her nose. "Luigi, maybe we had better put an end to this experiment."

"First let's solve the problem," I say, "by plugging 6 billion into our equation":

$$6,000,000,000 = 10e^{0.508t}$$

$$600,000,000 = e^{0.508t}$$

$$\ln(600,000,000) = 0.508t$$

$$t = \ln(600,000,000)/0.508 \text{ or } 39.7 \text{ hours.}$$

"In 39.7 hours, there will be 6 billion slime molds on the pizza."

"*Dio Mio!*" Rosario screams. " I don't want to be around here then!"

I chuckle. "No need to worry. I hope to stop the experiment before then."

■ ■ ■

Three days later, something goes terribly wrong. Despite my best efforts at eradication, the slime molds suddenly spread across the restaurant table and onto the floor. They are unstoppable. My customers scream as the carpet of mold begins to envelop the restaurant.

What do you do when something so absurd occurs before your very eyes, when the fabric of reality begins to tear? Oh, you can speculate how you would have fought the mold as you calmly sit reading these words, as you go about living your average life and addressing your mundane daily activities. But I don't think any of us had time to come to an intellectual understanding on that cold, cruel day when Fiona, Big Tony, Rosario, and I had to run for our lives. It was survival of the fittest. It was a day when mold became king.

EXERCISES..........................

1. Fungus in an old tomato sauce jar grows exponentially, and the initial number of spores has doubled in 6 hours. Even though we don't know the initial number of spores, we'd like to find out how many times the initial number will be present in 12 hours. If you solve this problem, Luigi might give you a new jar of sauce for your hard work.

2. Solve problem 1 if the initial number of spores has *tripled* in 6 hours.

3. For some strange reason, the number of olives in Luigi's big olive jar disappear exponentially (Figure 36). Perhaps customers are stealing them, or perhaps some fast-growing fungus is devouring the olives. Assume that 300 olives become 50 olives in 1 hour. About how many olives will remain after 2 hours?

4. Solve problem 3 assuming that 400 olives become 300 olives in 1 hour.

5. The population of Luigi's beloved homeland, Italy, was about 57 million in 1999. At this time, Luigi introduced his new garlic and oyster pizza, which caused the growth rate to rise exponentially with a growth constant of $k = 0.01$. At this rate, estimate Italy's population in the year 2029. You may use a calculator for this problem.

6. Solve problem 5 for the year 2040 and a growth constant of $k = 0.001$.

B.C. MANSFIELD

FIGURE 36. Disappearing olives.

7. Luigi has thousands of calzones stored in a warehouse. Unfortunately, alien rats with polka-dot teeth have gnawed their way into the warehouse, are having babies, and are eating the calzones (Figure 37). Assume that the number of calzones decays exponentially with a decay constant of k. The half-life T is defined to be the time interval after which half of the original number of calzones remains. Find the general relationship between k and T at the half-life point. Recall that $y = y_0 e^{kt}$.

8. In problem 7, what is the general relationship between k and T at the "third-life" point?

9. Assume that the half-life of the calzones in problem 7 is 100 days. Luigi enters the warehouse, screams, and notices that 10 percent of the original

B.C. MANSFIELD

FIGURE 37. Alien rat eating calzones.

calzones remain. You and Luigi must play the role of detectives. How many days before Luigi entered the warehouse did the rats enter the warehouse? You may use a calculator.

10. Assume that the half-life of the calzones in problem 9 is 200 days. Luigi enters the warehouse and notices that 5 percent of the original calzones remain. How many days before Luigi entered the warehouse did the rats enter the warehouse?

Calculus and Computers

In most sciences, one generation tears down what another has built, and what one has established another undoes. In mathematics alone each generation adds a new story to the old structure.

—Herman Henkel, in S. Gudder's *A Mathematical Journey*

Take a look at this," Big Tony says to me. He holds up a huge pizza and stares at me. Then he hands me a card with a question. There appear to be traces of slime mold on the edges of the card—the remnants of our recent experiment with exponential growth.

The lettering on the card is quite fancy. Perhaps Big Tony is trying to impress Fiona or me with the importance or difficulty of the question.

> IF THIS PIZZA WEIGHS 10 POUNDS
> PLUS HALF ITS OWN WEIGHT,
> HOW MUCH DOES IT WEIGH?

Can anyone help Fiona answer this odd question? If you think the question is difficult, you're not alone. If you think this is too easy, you may be incredibly brilliant and arrogant, but I bet that none of your friends can answer this within fifteen seconds. Try it on your friends. So far, none of my friends have solved it without a pencil and paper. If you're a teacher, have your students work on this problem. Allow them to use a pencil and paper. I'll give you the solution in the "Answers" section of this book.

"May I have the card?" Fiona asks. "I'd like to think about it more tonight."

Big Tony bows. "Certainly you may have the card."

"Big Tony," I say, "that's a fascinating question, but I'd like to finish our tour of calculus. I must apologize for not devoting large portions of our lessons to the use of computers in calculus. These days, inexpensive calculators can integrate functions. Computer software applications like Mathematica, Maple, and Mathcad can also solve calculus problems."

Fiona points her finger at me. "If you can use calculators and computers to solve calculus problems, then why learn calculus?"

"You can quickly make mistakes if you don't know the beginning concepts. It's difficult to use such computer tools and *understand* the results without first having a basic knowledge of calculus." I wait a few seconds. "I want you to think about our previous lessons on limits in which we saw how computers are good for understanding and estimating certain kinds of limits."

Fiona nods and points to the right pocket of her jeans. Embroidered on her pants in purple yarn are the words, I LOVE LIMITS.

I smile and nod. "Now, let's talk about derivatives. There are numerous ways of using a computer to calculate derivatives. One simple-minded approach uses the definition of derivative that we discussed in our early lessons." I write on the wall beneath a 1960s black-light poster of Isaac Newton

$$f'(t) = \lim_{\Delta t \to 0} \frac{f(t + \Delta t) - f(t)}{\Delta t}.$$

"We can rewrite this slightly to help us more clearly see how it's used to find the value of the derivative, $f'(a)$, at any number $x = a$ and when the derivative isn't necessarily calculated with respect to time":

$$f'(x) = \lim_{h \to 0} \frac{f(x+h) - f(x)}{h}.$$

"For example, we can compute this for a function $f(x) = x^3$ at the point $x = a = 3$. Because we are interested in this fraction in the limit as h gets increasingly small, we have to figure out how to produce smaller and smaller values of h using a computer program."

"I have an idea," says Fiona. "I took a class in BASIC programming. We might produce successively smaller values of h using the sequence produced by $h = 1/10^i$, where $i = 0, 1, 2, 3. \ldots$ This produces $h = 1, 0.1, 0.001, 0.0001$, and so forth."

"Excellent," I say. I pick up a pocket computer disguised as a salt shaker and show Fiona and Big Tony the LCD display. "Below is some BASIC code for computing the derivative of $f(x) = x^3$ when $x = a = 3$."

```
10 REM Compute derivative of function
20 A = 3
30 DEF FNY(X) = X*X*X
40 FOR I = 1 to 6
50     H = 1/10^I
60     Q = (FNY(A+H) - FNY(A))/H
70     PRINT H, Q
80 NEXT I
90 END
```

I point to the display. "Line 30 defines our function x^3. The FOR loop that starts in line 40 simply repeats six times the instructions between lines 40 and 80. In line 50, h gets smaller and smaller with each repetition of the FOR loop. In line 60, the program actually calculates the fraction with h as the denominator. In line 70, we print out the results."

"Let's give it a try," Fiona says. She presses a key marked "Go," and the computer spits out several pairs of numbers in the following table.

DERIVATIVE AT a = 3

h	Derivative Estimate
1.00000	37.0000
0.10000	27.9100
0.00100	27.0901
0.00010	27.0010
0.00001	27.0000

"You see," I say, "how the computer program can be used to estimate a value of 27 for the derivative when $x = a = 3$. As the values of h diminish, the solution appears to converge to 27."

"That was easy," Big Tony says with a growl. "Let's try to find the derivative at $x = 4$." He changes a line of the code from $A = 3$ to $A = 4$, and then he runs the BASIC program to produce another table of numbers below.

DERIVATIVE AT a = 4

h	Derivative Estimate
1.000000	61.0000
0.100000	49.2100
0.001000	48.1201
0.000100	48.0120
0.000010	48.0010
0.000001	48.0000

"Good," he says. "It seems to converge to 48."

I nod. "Of course, we could always take the derivative analytically and get $f(x) = x^3$, $f'(x) = 3x^2$. If we insert the value of $x = 3$, we get 27. If we insert the value of $x = 4$, we get 48."

Fiona puts her hands on her hips. "Luigi, why go through all the trouble to estimate the derivative using a computer program?"

"For one thing, many more difficult problems in the real world do not have simple analytical expressions for the derivative such as the one we had, $3x^2$. Computers can be very helpful in these cases. However, keep in mind that errors can creep into our computer calculations when h gets small and we start dividing expressions by increasingly smaller values. Notice that the more time we run the repetitive FOR loop in the BASIC program, the closer we approach the fraction 0/0. Computers only have a limited precision, so you always have to use these simple BASIC programs with care. Numerous procedures exist for improving the accuracy of these computer methods. They can be found in books like Sheldon Gordon's *Calculus and the Computer*."

A crowd is growing outside my restaurant. "Let's take a look at what's happening," I say.

Big Tony and Fiona follow me. Next to Trump Tower is a young woman serving sweet Turkish coffee followed by tiny tumblers of saruk, a strong anise-flavored liquor. Her husband is cooking small pieces of lamb and lobster threaded on a skewer, over an open fire. The food vendors are part of an annual multicultural fair. I tried some of the lamb just the other day, and it was delicious. The meat was usually marinated in lemon juice, olive oil, and spices and skewered with onions, bay leaves, and green peppers. Sometimes the man served it with rice and grilled vegetables.

"I'm starved," Fiona says, "but some of this food is a little—"

"Different," Big Tony says.

"Let's eat soon," I say. "But first I want to finish up my brief introduction to computers and calculus. Just as we can write programs to estimate the derivative, we can also create simple BASIC programs for estimating integrals. Recall from our previous lessons the definition of integration from $x = a$ to $x = b$." I use chalk to draw on the sidewalk

$$\int_a^b f(x)dx = F(b) - F(a) = [F(x)]_a^b.$$

"This can be recast slightly for easier implementation on a computer":

$$\int_a^b f(x)dx \approx (b-a)\sum_{i=0}^{n-1} \frac{f(x_i)}{n}.$$

Fiona hands me a pocket computer and I quickly type in a BASIC program. I hold up the device so that Fiona and Big Tony can see the program displayed on a small screen. "This BASIC program implements the integral for $f(x) = x^3$ from $x = 0$ to $x = 1$. We are essentially dividing the area under the curve into N thin rectangular areas to estimate the area under the curve."

```
10 REM Integration
20 DEF FNY(X) = X*X*X
30 A = 0
40 B = 1
50 N = 10
55 R = 0
60 H = (B-A)/N
70 FOR X = A TO B - H/2 STEP H
80     R = R + FNY(X)
90 NEXT X
100 R = R * H
110 PRINT "INTEGRATION ESTIMATE": R
```

"In line 60, you can see that the more rectangles N that we use, the smaller the rectangle's width h. The FOR loop from lines 70 to 90 repeatedly calculates the height of each rectangle, and we multiply by the constant width in line 100." I pause. "For this simple case, we know what the answer should be":

$$\int_0^1 x^3 dx = \left[\frac{x^4}{4} \right]_0^1 = \frac{1}{4} - 0 = \frac{1}{4}$$

"We can try running the program for increasing values of N, which leads to thinner rectangles and more accurate estimates. When N is 10,000 and the rectangle width is 0.0001, our estimate for the integration is 0.2499, which is just about right. As with the derivative, more sophisticated algorithms for estimating integrals exist."

INTEGRAL

N	Integral Estimate
2.0	0.0625
20.0	0.2256
40.0	0.2377
80.0	0.2438
200.0	0.2475
10000.0	0.2499

A few people are staring at my integral formula on the sidewalk. One man in a dirty hunting jacket stoops down to get a better look.

"Let's get some exercise," I say.

EXERCISES.....................

To conduct computational experiments relevant to the material presented in this chapter, you need a calculator or computer. For the programs that allow you to calculate derivatives and integrals, try experimenting with different values of h to assure yourself that the derivatives and integrals tend to be more accurate with smaller values of h. Of course, as mentioned, if you make h too small, your calculator or computer may start experiencing round-off errors due to the finite precision of numbers represented in computer programs. Can you find functions for which the programs do particularly poorly in estimating derivatives or integrals? Can you think of ways to improve the accuracy? Make plots of the derivative or integral values as a function of h.

There are no answers for this section, so feel free to experiment and observe your results. For example, find the derivative and integral of x^2 instead of x^3 by changing the function definitions in the programs to

```
DEF FNY(X) = X*X
```

In the integration program, experiment by extending the range of *x* values. You can do this by changing the current values in lines 30 and 40:

```
30 A = 0
40 B = 1
```

There is no end to the number of fascinating experiments you can try.

Multiple Integrals

> We have a habit in writing articles published in scientific journals to make the work as finished as possible, to cover up all the tracks, to not worry about the blind alleys or describe how you had the wrong idea first, and so on. So there isn't any place to publish, in a dignified manner, what you actually did in order to get to do the work.
>
> —Richard Feynman, Nobel lecture, 1966

"F iona, someday I want to read every calculus book in the world!"

"Luigi, that's a tall order! Amazon.com lists around thirty-five hundred calculus books."

I sigh. "Ah, perhaps you are right. The task may be too much for any person to accomplish. I like to imagine you and me on a soft, shaggy rug in front of a blazing fireplace doing calculus. Think how happy we would be."

Fiona smiles. "It is a tempting vision that we might one day pursue."

I take a deep breath. "When I was a boy poring through my father's mathematics books, I was always most impressed by those involving *multiple integrals*. Multiple integrals sent a delicious shiver up my spine. Here's an example from Cliff Pickover's book *Computers, Pattern, Chaos and Beauty*." I pull out a card and hand it to Fiona:

$$F(t,\omega) = \frac{1}{4\pi^2} \int\limits_{-\infty}^{+\infty} \int\limits_{-\infty}^{+\infty} \int\limits_{-\infty}^{+\infty} e^{-i\theta t - i\tau\omega + i\theta y} f(\theta,\tau)s^*\left(u - \frac{\tau}{2}\right)s\left(u + \frac{\tau}{2}\right)du\,d\tau\,d\theta.$$

"Ooh!" Fiona screams. "That is a good-looking equation."

"Isn't it a dandy fit for a king? I have always dreamed of writing a book that contained a splendid-looking formula like this with its multiple integral."

"What exactly is a multiple integral?" Fiona asks.

"Alas, our lessons are coming to a close, and I won't be dealing with multiple integrals. I'll save multiple integrals for next year's lessons on more advanced calculus. But to whet your appetite, consider this. Just as I taught you about how single integrals are good for calculating the area under a curve, double integrals are useful for computing volumes under a surface. Here's an example just to give you a flavor. Let's calculate the following." I write with a marker on Fiona's left hand

$$\int\limits_{2}^{3}\int\limits_{x}^{x^2} (3x + 2y)dy\,dx.$$

"The limits on the inner integral (x and x^2) are called the *inner limits*. The limits on the outer interval (3 and 2) are called the *outer limits*—not to be confused with the TV show."

Fiona studies the equation with exponentially increasing anticipation.

"When we find areas under curves, we are integrating over an interval on the x axis, for example, from $x = 0$ to $x = 10$. In our current multiple integral, the interval specifies a region in the xy-plane. In our case, the region is bounded below by the line $y = x$ and above by the curve $y = x^2$. The region is also bounded on the left by $x = 2$ and to the right by $x = 3$." I draw Figure 38 to help Fiona see the region for which we are finding the volume.

"In general, the double integral lets us find the volume under a surface $z = f(x,y)$ just like a single integral lets us find the area beneath the curve $y = f(x)$."

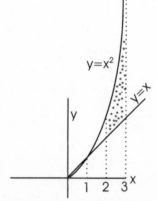

FIGURE 38. The region bounded by $y = x$ and $y = x^2$ and also by $x = 2$ and $x = 3$.

"What do you mean by 'under a surface'?" Fiona asks.

"We are finding the volume between the surface $z = (3x + 2y)$ and the xy-plane. Technically speaking, the double integral yields volume only when the function is always positive. Think of a chunk of parmesan cheese bounded in the xy-plane by the bounds in my drawing and above by the surface $z = (3x + 2y)$."

Fiona smiles, apparently enjoying the lesson.

"To solve the double integral, we first examine the inner integral . . ."

$$\int_{x}^{x^2} (3x + 2y)\,dy.$$

"When computing the inner integral, we don't treat x as a variable. Once we've done the integration, x returns to its variable status and we integrate with respect to x."

"Explain," Fiona says.

"The antiderivative of $3x$ with respect to y is $3xy$ when we treat $3x$ as a constant. An antiderivative of $2y$ is y^2. So far we have . . ."

$$\int_{2}^{3x^2}\int_{x}^{} (3x + 2y)\,dy\,dx = \int_{2}^{3} [3xy + y^2]_{y=x}^{y=x^2}\,dx.$$

"Next we want to evaluate the inner y integral from $y = x$ to $y = x^2$. We plug in the limits for y. Remember from previous lessons that the notation $[F(x)]_a^b$ is just a more compact way of writing $F(b) - F(a)$. So we get . . ."

$$\int_{2}^{3} [3xy + y^2]_{y=x}^{y=x^2}\,dx = \int_{2}^{3} \left\{ \left[3x(x^2) + (x^2)^2\right] - \left[3x(x) + x^2\right] \right\}dx$$

$$= \int_{2}^{3} (3x^3 + x^4 - 3x^2 - x^2)dx.$$

I begin to run out of space on Fiona's left arm so I start on her right. She doesn't seem to mind. Perhaps she thinks the marker uses washable ink.

"Remembering our antidifferentiation rule from previous lessons . . ."

$$\int u^r du = \frac{u^{r+1}}{r+1}.$$

"This equals . . ."

$$\int_{2}^{3}(3x^3 + x^4 - 4x^2)dx = \left[\frac{3x^4}{4} + \frac{x^5}{5} - \frac{4x^3}{3}\right]_2^3$$

$$= \left[\frac{3(3^4)}{4} + \frac{(3^5)}{5} - \frac{4(3^3)}{3}\right] - \left[\frac{3(2^4)}{4} + \frac{(2^5)}{5} - \frac{4(2^3)}{3}\right].$$

"This is equal to . . ."

$$\left[\frac{3(81)}{4} + \frac{(243)}{5} - \frac{4 \times 27}{3}\right] - \left[\frac{3(16)}{4} + \frac{(32)}{5} - \frac{4 \times 8}{3}\right]$$

$$= 73.35 - 7.733 = 65.616.$$

"In summary, we first evaluate a multiple integral with the inside integral (with respect to y in this case), treating the outside variable x as a constant. Once we solve the inside integral, we next integrate this with respect to the outside variable (x in this case)."

Fiona is deep in thought. "What sense do these integrals make in the real world? I don't just mean multiple integrals. I mean *any* integrals. Of what use are they?"

"Think of integrals as a fancy way to multiply. We can find the area of a rectangle by multiplying width by height. Integrals allow us to find areas when the height of the object is not flat or straight."

Fiona nods. "Integrals are good for finding areas—or volumes in the case of double integrals—but what else?"

I put the marker in my pocket and take a deep breath. "Integrals are good for finding solutions whenever products are involved. We immediately think of area because the area of a rectangle is the width times the height. We can see how integrals can help with that. But they are also good for finding other kinds of products, for example, the amount of work done, which is force times distance, or any quantity involving a multiplication. Similarly, multiple integrals are useful when several variables are being multiplied."

I open the window and gaze up at the incredible lamp of stars. A cool breeze blows through the window, and in the distance I hear the horn of a tug boat. We are silent for a minute, enjoying our calm surroundings.

I turn to Fiona. "Multiple integrals are everywhere these days. For example, a quick scan of the web using the Google search engine reveals several entries with the words *quadruple integrals*. Applications on the web range from calculations involving integrated circuits to studying temperatures in truck cabins. Fiona, you should try searching the web for more applications. Statistical optics papers are full of quadruple integrals. Crystallographers also use multiple integrals." I show Fiona the results of my web search. "Look here, *quintuple* integrals can be found in papers discussing special kinds of gases called plasmas."

Fiona looks down at the ground. She seems nervous.

"What?" I ask.

"Luigi, I don't mean to pry into your personal life."

"Yes?"

"If you are a genius in calculus, why are you a pizza chef?" Fiona asks.

I take a deep breath. "I believe that the true expression of genius occurs when someone can hold more than one idea in his or her head at the same time. I like thinking about calculus and pizza. It makes me happy. It's a good life."

Fiona takes my hand in hers and gives it a squeeze. Someday we will spend our days on a secluded beach doing calculus, proving theorems, and occasionally forgetting about the rest of the world. But for now it is time to leave calculus for a few hours. Life must go on. Calculus is wonderful, but I have all the rest of mathematics to explore.

The next day I spend time with Fiona at a local coffeehouse with spruce branches hanging over the entrance. The tiramisu is accompanied by bass, piano, and guitar music. Big Tony plays the guitar. On his shirt are the words I LOVE MULTIPLE INTEGRALS, and his big guitar is in the shape of a

$$\int.$$

EXERCISES

Relax. The story and lessons are finished. No exercises today. If you have gotten this far in the book, go out and reward yourself with a pizza. *Ciao.*

16

Calculus in a Nutshell

The union of the mathematician with the poet, fervor with measure, passion with correctness, this surely is the ideal.

—William James, *Collected Essays*

DERIVATIVES (CHAPTERS 2 THROUGH 4)

Derivative Definition

$$f'(t) = \lim_{\Delta t \to 0} \frac{f(t + \Delta t) - f(t)}{\Delta t}$$

Some Simple Derivative Rules

Rule Name	Formula	Example
Power rule	$f(t) = t^n$ $f'(t) = nt^{n-1}$ (n is a real number)	$f(t) = t^4; f'(t) = 4t^3$
Multiplier rule	$(af)' = af'$ (a is a constant multiplier)	$(a \times 3t^2)' = a(3t^2)'$
Sum rule	$(f + g)' = f' + g'$	$(3t^2 - 2t^4)' = (3t^2)' - (2t^4)'$

Rule Name	Formula
Product rule	$(fg)' = f'g + fg'$
Quotient rule	$\left(\dfrac{f}{g}\right)' = \dfrac{f'g - fg'}{g^2}$

CHAIN RULE AND IMPLICIT DIFFERENTIATION (CHAPTER 5)

Chain Rule

$$\frac{dy}{dx} = \frac{dy}{du} \times \frac{du}{dx}$$

$$\frac{d}{dx} f(g(x)) = f'[g(x)]g'(x)$$

Derivatives for Sin and Cos

$$(\sin x)' = \cos x$$

$$(\cos x)' = -\sin x$$

$$(\sin u)' = (\cos u)u'$$

$$(\cos u)' = -(\sin u)u'$$

MAXIMA AND MINIMA (CHAPTER 6)

Critical Point at x_c

Second Derivative	Type of Critical Point	Memory Aid
$f''(x_c) > 0$	Local minimum at x_c	⇑⇓
$f''(x_c) < 0$	Local maximum at x_c	⇓⇑

Concavity and Slope

First Derivative	Second Derivative	Classification	Pictorial
$f'(x) < 0$	$f''(x) > 0$	Curve decreasing, concave up	Left side of ⌣
$f'(x) > 0$	$f''(x) > 0$	Curve increasing, concave up	Right side of ⌣
$f'(x) < 0$	$f''(x) < 0$	Curve decreasing, concave down	Right side of ⌢
$f'(x) > 0$	$f''(x) < 0$	Curve increasing, concave down	Left side of ⌢

EXPONENTIALS AND LOGARITHMS (CHAPTER 8)

Rules for Manipulating Logarithms

Rule Name	Logarithm Rule
Product	$\log_b xy = \log_b x + \log_b y$
Quotient	$\log_b(x/y) = \log_b x - \log_b y$
Reciprocal	$\log_b(1/x) = -\log_b x$
Power	$\log_b x^p = p\log_b x$
Base change	$\log_c x = \log_b x / \log_b c$
Addition	None. Note: $\log_b(x + y) \neq \log_b x + \log_b y$

Exponential Rules

Rule Name for Derivative	Exponential Rule
Exponential	$(e^x)' = e^x$
Exponential (chain)	$(e^u)' = e^u u'$
Exponential base b	$(b^x)' = (\ln b)b^x$
Exponential base b (chain)	$(b^u)' = (\ln b)b^u u'$
Logarithm	$(\ln x)' = 1/x$
Logarithm (chain)	$(\ln u)' = (1/u)u'$
Logarithm base b	$(\log_b x)' = (1/\ln b)(1/x)$
Logarithm base b (chain)	$(\log_b u)' = (1/\ln b)(1/u)u'$

LIMITS AND CONTINUITY (CHAPTER 9)

L'Hôpital's Rule

$$\lim_{x \to a} \frac{f(x)}{g(x)} = \lim_{x \to a} \frac{f'(x)}{g'(x)}$$

Continuity at a

1. $f(a)$ is defined

2. $\lim_{x \to a} f(x)$ exists

3. $\lim_{x \to a} f(x) = f(a)$

INTEGRATION (CHAPTER 11)

Antiderivatives

Function	Antiderivative
0	C
a	$ax + C$
x	$\frac{1}{2}x^2 + C$
x^r	$\dfrac{x^{r+1}}{r+1} + C \ (r \neq -1)$
x^{-1}	$\ln |x| + C \ (x \neq 0)$
$\cos ax$	$\dfrac{1}{a}\sin ax + C$
$\sin ax$	$-\dfrac{1}{a}\cos ax + C$
e^{ax}	$\dfrac{1}{a}e^{ax} + C$
$\dfrac{1}{\sqrt{a^2 - x^2}}$	$\arcsin \dfrac{x}{a} + C$
$\dfrac{1}{a^2 + x^2}$	$\dfrac{1}{a}\arctan \dfrac{x}{a} + C$

Rules for Integration

Rule	Notation
Sign switch	$\int_a^b f(x)dx = -\int_b^a f(x)dx$
Sum	$\int_a^b \left[f(x)+g(x)\right]dx = \int_a^b f(x)dx + \int_a^b g(x)dx$
Constant multiplier	$\int_a^b cf(x)dx = c\int_a^b f(x)dx$
Intervals	$\int_a^c f(x)dx = \int_a^b f(x)dx + \int_b^c f(x)dx$
Area between two curves	$\int_a^b \left[f(x)-g(x)\right]dx$

Antiderivatives

Indefinite Integrals	Antiderivative
$\int u^r du$	$\dfrac{u^{r+1}}{r+1} + C \ (r \neq 1)$
$\int u^{-1} du$	$\ln\lvert u \rvert + C \ (u \neq 0)$
$\int \cos u\, du$	$\sin u + C$
$\int \sin u\, du$	$-\cos u + C$
$\int e^u du$	$e^u + C$
$\int \dfrac{du}{\sqrt{a^2 - u^2}}$	$\arcsin \dfrac{u}{a} + C$
$\int \dfrac{du}{a^2 + u^2}$	$\dfrac{1}{a}\arctan \dfrac{u}{a} + C$

LOGARITHMIC DIFFERENTIATION, INTEGRATION BY PARTS, TRIGONOMETRIC SUBSTITUTION, AND PARTIAL FRACTIONS (CHAPTER 12)

Integration by Parts

$$\int u\, dv = uv - \int v\, du$$

EXPONENTIAL GROWTH AND DECAY (CHAPTER 13)

Exponential Growth and Decay

$$\frac{dN}{dt} = kN$$

$$N(t) = N_0 e^{kt}$$

17

Luigi's Mind-Boggling Workout Routine

Calculus need not be made easy; it is easy already. However, it is a subject which cannot be mastered by sheer memory. On the contrary, the student must accustom himself to memorizing as few formulas as possible and to reasoning out the correct attack in any situation.

—G. M. Petersen and R. F Graesser,
Differential and Integral Calculus

Learning is often best done through examples. This section contains some solved problems and applications from several different areas in calculus. If you succeed in going through Luigi's workout, your mind will be expanded and all your desires satiated. Treat yourself to a calzone after you have understood all the problems in this section.

Note that the phrase "mind boggling" in the chapter title is an exaggeration. By this point in the book, you should be able to handle these problems. Your mind won't be boggled. It will just be exercised.

EXERCISES.........................

1. Luigi has just finished making tomato sauce from tomatoes, salt, and fresh basil leaves. Now he wants to manufacture a can for the tomato sauce. The can should contain a volume of 100 cubic inches. Luigi also wants to use as little material as possible for the can itself. What is the radius of the optimal tomato sauce can? (Hint: The volume of a cylinder is $\pi r^2 h$.)

2. As in problem 1, Luigi wants to manufacture a can of tomato sauce that uses the least amount of material possible. This time he wants to impress Fiona by creating a can that will have a volume of 1,000 cubic inches! What is the radius of the can?

3. Luigi has a number of pizzas in his oven. The number of pizzas plus 10 pepperoni is 2 less than 5 times the number of pizzas. If you wish, denote the number of pizzas by P_1 and the number of pepperoni by P_2. How many pizzas does Luigi have?

4. Luigi has a number of olives on his pizza. The number of olives plus 14 anchovies is 2 less than 5 times the number of olives. If you wish, denote the number of olives by O and the number of anchovies by A. How many olives does Luigi have?

5. Here is a limit that for some strange reason describes the number of people in Luigi's restaurant right as you read these words. Find the limit

$$\lim_{x \to \infty} \frac{15x^3 - 2x + 10}{5x^3 + \pi}.$$

6. Find the limit

$$\lim_{x \to \infty} \frac{20x^4 - 25x + \beta}{5x^4 + \pi^2},$$

where

$$\beta = \frac{d}{dx}\left(\frac{1}{\pi}x + e^2\right).$$

7. Find the derivative of

$$\frac{1}{(\pi x^3 + 3x^2 + 20x + 5)^4}.$$

8. Find the derivative of

$$\frac{1}{(ex^3 + 2x^2 + 10x + 5)^5}.$$

9. Find

$$\int \frac{e}{\sqrt{x - \pi}} dx.$$

10. Find

$$\int \frac{a}{\pi\sqrt{x - \pi}} dx + \int 2x^3 dx,$$

where

$$a = \frac{d}{dx}\left(\frac{1}{e}x + \pi e^3\right).$$

Conclusion

Hello. Luigi here. I don't like books with long-winded conclusions. Let me wrap up by saying Fiona and I eventually became engaged. Each day after working in my pizza shop, we relax in my apartment above the shop and do calculus into the wee hours of the night. It's a nice life.

Before saying good-bye, I wanted to tell you a little about Newton's fundamental anagram of calculus:

THE FUNDAMENTAL ANAGRAM OF CALCULUS

6accdae13eff7i3l9n4o4qrr4s8t12ux

Isaac Newton and Gottfried Wilhelm von Leibniz had a feud over who really discovered calculus. One incident from 1677 is particularly interesting and relevant. At this time, Newton was answering several of von Leibniz's questions about infinite series. In his letter, Newton came close to revealing his fluxional method—that is, Newton's own version of calculus. However, instead of revealing the method, Newton concealed it in the form of an anagram. Perhaps Newton used the anagram because he didn't want von Leibniz to scoop him and because Newton wanted a way to demonstrate he had actually known about calculus should he have to prove this at a later date. Whatever the reason, Newton was not ready to give a full explanation.

After Newton described his methods of tangents and finding maxima and minima, he wrote to von Leibniz: "The foundations of these operations are evident enough, in fact; but because I cannot proceed with the explanation of it now, I have preferred to conceal it thus: 6accdae13eff7i3l9n4o4qrr4s8t12ux. On this foundation I have also tried to simplify the theories which concern the squaring of curves, and I have arrived at certain general Theorems" (Richard Westfall, *Never at Rest*. New York: Cambridge University Press, 1980, p. 265; see also, www.mathpages.com/home/kmath414.htm).

The anagram expresses, in Newton's terminology and in Latin, the fundamental theorem of calculus: *"Data aequatione quotcunque fluentes quantitates involvente, fluxiones invenire; et vice versa,"* which means "Given an equation involving any number of fluent quantities to find the fluxions, and vice versa." (*Fluxion* was Newton's term for derivative.) The numbers in the anagram count the number of letters in the sentence. For example "6a" corresponds to the six occurences of letter *a* in the sentence. There are thirteen occurences of letter *e*, and so forth.

Interestingly, neither von Leibniz nor Newton had published papers on calculus at the time this letter was exchanged, although both probably were well versed in the subject. Thus, if Newton had avoided using the anagram and had sent von Leibniz an explicit statement about his calculus, Newton might have established his superior knowledge beyond doubt. Instead, Newton's secret anagrams caused him to lose his possible claim to calculus and led to heated arguments that plagued his and von Leibniz's lives for years to come.

▪ ▪ ▪

A Final Reward: I am happy to send you a handsome certificate attesting to your mathematical prowess if you send me the correct solution to the following. Write to Luigi, c/o Dr. Cliff Pickover, P.O. Box 549, Millwood, New York 10546-0549. Please show your work:

$$\int\left[\left(\frac{e}{x^3}-\frac{e}{x^5}\right)\times\pi+\cos x\right]dx.$$

Answers

> What I am going to tell you about is what we teach our physics students in the third or fourth year of graduate school. . . . It is my task to convince you not to turn away because you don't understand it. You see my physics students don't understand it. . . . That is because I don't understand it. Nobody does.
>
> —Richard Feynman, *QED: The Strange Theory of Light and Matter*

2. Derivative Definitions and Rules

1. Given $f(t) = 3t^2 + 4t^3$, we can find the derivative $f'(t) = 6t + 12t^2$. When $t = 100$, the derivative is 120,600, which means the meatball is traveling at 120,600 miles per second. Of course, this is outrageously fast, considering the speed of light is 186,000 miles per second in a vacuum! In real life, Rosario's meatball (or anything else he tried to accelerate to this speed) would have disintegrated long before it reached this speed.

2. Given

$$T(d) = 160 - \frac{1}{2}d^2,$$

we can find the temperature of the pizza by evaluating the function for T at one-half inch. So $T(1/2) = 159\frac{7}{8}$ degrees, which is not a big change from the center temperature. To find the rate of temperature change, find the derivative of $T(d)$, which is $T'(d) = -d$. At one-half inch from the center, the rate of temperature change is $T'(1/2) = -\frac{1}{2}$ degree/inch.

3a. Given

$$f(t) = \frac{1}{10,000,000}(5t^2 + 3)^2,$$

we can estimate the number of people in the restaurant 60 minutes after 7 P.M. without even taking a derivative. All we need to do is substitute $t = 60$ in the formula, which produces

$$F(60) = \frac{1}{10,000,000}(5 \times 60^2 + 3)^2 = \frac{1}{10,000,000}(5 \times 3,600 + 3)^2,$$
$$= 324,108,009/10,000,000,$$

which is about 32 people.

3b. The problem asks that we determine the rate at which people enter the restaurant at 8 o'clock. Therefore, we want to find the derivative of

$$f(t) = \frac{1}{10,000,000}(5t^2 + 3)^2.$$

First let's expand $(5t^2 + 3)^2$ so that we can use the power rule. In Chapter 1, we encountered the binomial formula, which says $(a + b)^2 = a^2 + 2ab + b^2$. Here we can let $a = 5t^2$ and $b = 3$. Thus, the expression becomes $(5t^2)^2 + 2 \times 5t^2 \times 3 + 3^2 = 25t^4 + 30t^2 + 9$. The next step is to take the derivative of each term, and let's not forget to consider the fractional multiplier, which we have so far left out of the computation. Thus, we get

$$f'(t) = \frac{1}{10,000,000}(100t^3 + 60t).$$

So, at 8 P.M. (or $t = 60$) we have $21,603,600/10,000,000$, or about 2 people entering the restaurant per minute. That's a nice rate, and hopefully Luigi will sell a lot of pizzas.

We wouldn't expect a simple equation like this to be accurate for extended periods of time because $f(t)$ gets larger and larger through time, and we don't expect Luigi's restaurant to have a huge capacity.

So far we have computed the *rate* of people entering the restaurant. The actual number of people in the restaurant at 8 o'clock at night can be determined simply by substituting a value of $t = 60$ in the original equation. This yields

$$f(60) = \frac{1}{10,000,000}(5 \times 60^2 + 3)^2,$$

which is about 32 people.

4. Given $y = 3x^2 - 3x + 2$, the derivative y' is $6x - 3$.

5. Given $y = 4x^3 - \frac{1}{2}x + 20$, the derivative y' is $12x^2 - \frac{1}{2}$.

6. Remember, the rate of change of a constant is zero. So, for $y = 1,000,000$, the derivative $y' = 0$.

7. We are asked to compute the derivative $y' = dy/dx$ for $y = \pi^2$. The derivative of a constant is zero. Therefore, $y' = 0$.

8. Given

$$y = \frac{1}{2}\pi^2 x - \frac{\pi^3}{3},$$

the derivative is

$$y' = \frac{1}{2}\pi^2.$$

9. We are asked to compute the derivative

$$y' = dy/dx \text{ for } y = \frac{\pi^3 x}{3} - \frac{ax^3}{2},$$

where $a = 4$. Thus,

$$y' = \frac{\pi^3}{3} - \frac{3ax^2}{2} = \frac{\pi^3}{3} - 6x^2.$$

10. Given

$$y = \frac{\pi^3 x^2}{3} - \frac{ax^4}{2},$$

we are to find the derivative where $a = 4$. Therefore, the derivative is

$$y' = \frac{2\pi^3}{3}x - 2ax^3 = \frac{2\pi^3}{3}x - 8x^3.$$

3. Derivatives and Slopes

1. We are given the equation $y = x^3 - 1$. First we find the derivative, which is $y' = 3x^2$. This means that the slope of the tangent line at $x = 2$ is 12. The slope of the tangent line at $x = -3$ is 27. This is the solution to part (a). For part (b), note that the curve is steeper at $x = -3$ because the slope of 27 is greater than the slope of 12. For part (c), the equation for the line tangent to $x = 2$ is $y = 12x + b$. Let's solve for b. We know that $x = 2$ corresponds to a point on the tangent line. At $x = 2$, $y = 7$. Using these values in the equation $y = 12x + b$, we find the equation of the tangent line is $y = 12x - 17$.

2. The equation of the line tangent to the curve $y = x^2 + 10$ at $x = 2$ is found by first determining the slope at that point. The derivative is $y' = 2x$. Therefore, the slope at $x = 2$ is 4. The point on the curve at $x = 2$ is $y = (2)^2 + 10 = 14$. Using $y = mx + b$ with $m = 4$ and $(x, y) = (2, 14)$, the equation of the line tangent to the curve $y = x^2 + 10$ at $x = 2$ is $y = 4x + 6$. The slope at $x = -2$ is -4, while the slope at $x = 1$ is 2. The curve is steeper at $x = -2$ because the change in y is much greater than the change in x as compared to the slope when $x = 1$.

3. Given the curve defined by $y = x^2 - x + 6$, the derivative is $y' = 2x - 1$. The slope is zero at $x = \frac{1}{2}$. Therefore, the curve is least steep at $(\frac{1}{2}, 5\frac{3}{4})$. Notice that we obtained the value $5\frac{3}{4}$ by inserting $\frac{1}{2}$ into the original formula, $y = (\frac{1}{2})^2 - \frac{1}{2} + 6$.

4. Given $y = -x^2 + x + 10$, the derivative is $y' = -2x + 1$. When the slope is zero, $x = \frac{1}{2}$, and

$$y = -\left(\frac{1}{2}\right)^2 + \frac{1}{2} + 10 = \frac{41}{4}.$$

So, the meatball would settle at the point $(\frac{1}{2}, 10\frac{1}{4})$.

5. We must calculate the derivative of $y = \pi x^5 - x$, recalling that π may be treated like any other constant. The derivative is $y' = 5\pi x^4 - 1$. Therefore, the slope at $x = 0$ is -1.

6. Given $y = \pi x^6 - x^2$, the derivative is $y' = 6\pi x^5 - 2x$. When $x = 0$, the slope of the tangent is zero.

7. We must calculate the derivative of

$$y = \frac{\pi\sqrt{x}}{2},$$

which can be rewritten as

$$y = \frac{\pi x^{\frac{1}{2}}}{2}.$$

Thus,

$$y' = \frac{\pi\frac{1}{2}x^{-\frac{1}{2}}}{2} = \frac{\pi x^{-\frac{1}{2}}}{4} = \frac{\pi}{4\sqrt{x}}.$$

(To help you manipulate these types of expressions, recall the rule of exponents that states $x^{-a} = 1/x^a$.) At $x = 1$, the slope is $\pi/4$.

8. We can rewrite

$$y = \frac{\pi\sqrt{x}}{2} + 3x^2$$

as

$$y = \frac{\pi(x)^{\frac{1}{2}}}{2} + 3x^2.$$

The derivative is

$$y' = \frac{\pi(x)^{-\frac{1}{2}}}{4} + 6x = \frac{\pi}{4\sqrt{x}} + 6x.$$

At $x = 1$, the slope of the tangent is $\pi/4 + 6$.

9. $y = (x + 2)^2 = (x + 2)(x + 2) = x^2 + 4x + 4$. $y' = 2x + 4$. Thus, the slope at $x = 0$ is 4.

10. Given $y = (x + 3)^2$, multiply the binomial in order to use the product/multiplier rules. Therefore, $y = x^2 + 6x + 9$ and the derivative is $y' = 2x + 6$. The slope of the tangent at $x = 0$ is 6.

4. Rules for Products and Quotients

1. Yes, you can help Big Tony if you use the quotient rule. We can let $f = x$ and $g = x^3 - 1$. The quotient rule states that

$$\left(\frac{f}{g}\right)' = \frac{f'g - fg'}{g^2}.$$

Thus,

$$\left[\frac{x}{x^3 - 1}\right]' = \frac{(1)(x^3 - 1) - x(3x^2)}{(x^3 - 1)^2} = \frac{x^3 - 3x^3 - 1}{(x^3 - 1)^2} = \frac{-2x^3 - 1}{(x^3 - 1)^2}.$$

2. Given

$$\frac{x^2 + 2x + 2}{x^2 + 1},$$

let $f = x^2 + 2x + 2$ and $g = x^2 + 1$. The derivatives are $f' = 2x + 2$ and $g' = 2x$. Use the quotient rule to obtain the derivative

$$\frac{(2x + 2)(x^2 + 1) - (x^2 + 2x + 2)(2x)}{(x^2 + 1)^2} = \frac{-2(x^2 + x - 1)}{(x^2 + 1)^2}.$$

3. We use the product rule to get

$$2x(x^2 + 2) + (x^2 + 1)(2x) \text{ or } 2x\left\{(x^2 + 2) + (x^2 + 1)\right\} = 2x(2x^2 + 3).$$

The slope at $x = 0$ is zero.

4. Given $y = (\frac{1}{2}x^3 + 1)(x^4 + x^2 - 100)$, let $f = \frac{1}{2}x^3 + 1$ and $g = x^4 + x^2 - 100$. Therefore, the derivatives are $f' = \frac{3}{2}x^2$ and $g' = 4x^3 + 2x$. Using the product rule, we obtain

$$y' = \left(\frac{3}{2}x^2\right)(x^4 + x^2 - 100) + \left(\frac{1}{2}x^3 + 1\right)(4x^3 + 2x)$$

$$= \frac{7}{2}x^6 + \frac{5}{2}x^4 + 4x^3 - 150x^2 + 2x.$$

The slope of the curve at $x = 0$ is zero.

5. We are given

$$y = \left(\sqrt{x} + 1\right)(x^2 + 2x + 2),$$

which is the same as

$$y = \left(x^{\frac{1}{2}} + 1\right)(x^2 + 2x + 2).$$

The product rules gives us

$$y' = \left(\frac{1}{2}x^{-\frac{1}{2}}\right)(x^2 + 2x + 2) + \left(x^{\frac{1}{2}} + 1\right)(2x + 2).$$

6. Given

$$y = \left(2\sqrt{x} + 1\right)(x^2 + 2x - 2),$$

let $f = 2x^{\frac{1}{2}} + 1$ and $g = x^2 + 2x - 2$. The derivatives are $f' = x^{-\frac{1}{2}}$ and $g' = 2x + 2$. Using the product rule, we obtain

$$y' = \left(x^{-\frac{1}{2}}\right)(x^2 + 2x - 2) + \left(2x^{\frac{1}{2}} + 1\right)(2x + 2)$$

$$= 5x^{\frac{3}{2}} + 2x + 6x^{\frac{1}{2}} + 2 - 2x^{-\frac{1}{2}}$$

$$= 5x\sqrt{x} + 2x + 6\sqrt{x} + 2 - \frac{2}{\sqrt{x}}.$$

7. We are given

$$y = \frac{x}{\pi}.$$

Of course, we hardly needed to use the quotient rule for such a simple function. But we can use the rule as a learning aid, and then we can examine the result to determine if the answer makes sense: $f = x$ and $g = \pi$: $f' = 1$ and $g' = 0$. Thus,

$$\left(\frac{f}{g}\right)' = \frac{1 \times \pi - x \times 0}{\pi^2} = \frac{1}{\pi}.$$

8. Given

$$y = \frac{x^2}{\pi},$$

let $f = x^2$ and $g = \pi$. The derivatives are $f' = 2x$ and $g' = 0$. Using the quotient rule, we obtain

$$y' = \frac{2x(\pi) - x^2(0)}{\pi^2} = \frac{2x}{\pi}.$$

9. We are given

$$\frac{1}{x}\sqrt{x}$$

and told to use the product rule to find the derivative. Let $f = 1/x$ and $g = \sqrt{x}$:

$$\left(\frac{1}{x}\sqrt{x}\right)' = -x^{-2}x^{\frac{1}{2}} + \frac{1}{x}\frac{1}{2}x^{-\frac{1}{2}}.$$

Let's simplify this, remembering the general rule of exponents that says $x^a x^b = x^{a+b}$:

$$-x^{2}x^{\frac{1}{2}} + \frac{1}{x}\frac{1}{2}x^{-\frac{1}{2}} = -x^{-\frac{3}{2}} + \frac{1}{2}x^{-\frac{3}{2}} = -\frac{1}{2}x^{-\frac{3}{2}}.$$

Of course, we could have avoided the product rule if we had recast the original form as

$$x^{-1}x^{\frac{1}{2}} = x^{-\frac{1}{2}}$$

and then taken the derivative, but this would have given us less practice with the product rule. You can also try to solve this using the quotient rule.

10. Given

$$\frac{\pi}{2x}\sqrt{x},$$

let

$$f = \frac{\pi}{2x},$$

which can be written as

$$f = \frac{\pi}{2}x^{-1},$$

and

$$g = \sqrt{x},$$

which can be written as

$$g = x^{\frac{1}{2}}.$$

The derivatives are

$$f' = \frac{-\pi}{2}x^{-2}$$

and

$$g' = \tfrac{1}{2}x^{-\frac{1}{2}}.$$

Therefore,

$$\left(\frac{\pi}{2x}\sqrt{x}\right)' = \left(\frac{-\pi}{2}x^{-2}\right)\left(\sqrt{x}\right) + \left(\frac{\pi}{2x}\right)\left(\frac{1}{2}x^{-\frac{1}{2}}\right) = \frac{-\pi}{4}x^{-\frac{3}{2}} = \frac{-\pi}{4x\sqrt{x}}.$$

We could have simplified

$$\frac{\pi}{2x}\sqrt{x} \text{ to } \frac{\pi}{2}x^{-\frac{1}{2}}$$

for which the derivative is

$$\frac{-\pi}{4}x^{-\frac{3}{2}} = \frac{-\pi}{4x\sqrt{x}}.$$

5. Chain Rule and Implicit Differentiation

1. We want to use the chain rule to compute the derivative dy/dx of

$$\left(\frac{x^4}{4}+4\right)^4.$$

Let's recast this by expressing the problem in terms of u^4, where

$$u = \left(\frac{x^4}{4}+4\right).$$

To find the derivative, we first compute the derivative of u^4, which is $4u^3$. (Another way of expressing this is $dy/du = 4u^3$.) We also need to take the derivative of u with respect to x—that is, du/dx. This yields $4x^3/4$ or x^3. Using the chain rule,

$$\frac{dy}{dx} = \frac{dy}{du} \times \frac{du}{dx},$$

we obtain

$$\frac{dy}{dx} = 4u^3 \times x^3.$$

Substituting the actual value for u, we get

$$4\left(\frac{x^4}{4}+4\right)^3 \times x^3.$$

The slope at $x = 1$ is $4(17/4)^3 = 17^3/4^2 = 17^3/16$.

2. Use the chain rule to find the derivative of

$$y = \left(\frac{x^3}{\pi} + \pi\right)^3 - \left(\frac{x^2}{\pi} + \pi\right)^2.$$

The derivative is

$$y' = \frac{9}{\pi}x^2\left(\frac{x^3}{\pi} + \pi\right)^2 - \frac{4x}{\pi}\left(\frac{x^2}{\pi} + \pi\right).$$

To find the slope, evaluate $y'(1)$, which is

$$y'(1) = \frac{9}{\pi}\left(\frac{1+\pi^2}{\pi}\right)^2 - \frac{4}{\pi}\left(\frac{1+\pi^2}{\pi}\right)$$

$$= \left(\frac{1+\pi^2}{\pi^2}\right)\left(\frac{9}{\pi}(\pi^2 + 1) - 4\right) = 29.88.$$

3. We want to use implicit differentiation to compute the derivative dy/dx of $x^3y - (y + x)^4 + 1 = 0$. First we take the derivative of each term of the expression, a process that can be written as

$$\frac{d}{dx}(x^3y) - \frac{d}{dx}(y+x)^4 + \frac{d}{dx}1 = \frac{d}{dx}0.$$

We can use the product rule to evaluate the first term. The product rule tells us that $(fg)' = f'g + fg'$, and in this case $f = x^3$ and $g = y$. We can use the chain rule to evaluate the second term. This gives us

$$3x^2y + x^3\frac{dy}{dx} - 4(y+x)^3\left(\frac{dy}{dx} + 1\right) + 0 = 0.$$

Next we regroup to solve for dy/dx:

$$x^3 \frac{dy}{dx} - 4(y+x)^3 \left(\frac{dy}{dx} + 1 \right) = -3x^2 y$$

$$\frac{dy}{dx} \times \left[x^3 - 4(y+x)^3 \right] - 4(y+x)^3 = -3x^2 y$$

$$\frac{dy}{dx} \times \left[x^3 - 4(y+x)^3 \right] = -3x^2 y + 4(y+x)^3$$

$$\frac{dy}{dx} = \frac{-3x^2 y + 4(y+x)^3}{\left(x^3 - 4(y+x)^3 \right)}.$$

The slope at $(0, 1)$ is

$$\frac{4}{-4(1)^3} = -1.$$

4. Given $ax + by = 0$, take the derivative of each term of the equation

$$\frac{d}{dx}(ax) + \frac{d}{dx}(by) = \frac{d}{dx}(0).$$

We obtain

$$a + b\frac{dy}{dx} = 0 \text{ or } \frac{dy}{dx} = \frac{-a}{b}.$$

The slope at $(-b, a)$ will be

$$\frac{-a}{b}.$$

Note that the slope is a constant since this is the equation of a straight line.

5. We want to use the extended power rule to compute

$$\left(\frac{\pi}{x^2 + 1} \right)'.$$

Let's do it as follows:

$$\left(\frac{\pi}{x^2+1}\right)' = \pi\left[(x^2+1)^{-1}\right]' = -\pi(x^2+1)^{-2}2x = -\frac{2\pi x}{(x^2+1)^2}.$$

The slope at $x = 1$ is

$$-\frac{2\pi}{4} = -\frac{\pi}{2}.$$

6. The expression

$$\frac{\pi}{\sqrt{x^2+\pi}} + \sqrt{\sqrt{x}}$$

is the same as

$$\frac{\pi}{(x^2+\pi)^{\frac{1}{2}}} + x^{\frac{1}{4}} = \pi(x^2+\pi)^{-\frac{1}{4}} + x^{\frac{1}{4}}.$$

Using the extended power law, we get the derivative

$$\frac{-1}{2}\pi(x^2+\pi)^{-\frac{3}{2}}2x + \frac{1}{4}x^{-\frac{3}{4}} = \frac{-x\pi}{(x^2+\pi)^{\frac{3}{2}}} + \frac{1}{4x^{\frac{3}{4}}}$$

$$= \frac{(-x\pi)(4x^{\frac{3}{4}})(x^2+\pi)^{\frac{3}{2}}}{4x^{\frac{3}{4}}(x^2+\pi)^{\frac{3}{2}}} = \frac{-4x^{\frac{7}{4}}\pi + (x^2+\pi)^{\frac{3}{2}}}{4x^{\frac{3}{4}}(x^2+\pi)^{\frac{3}{2}}}.$$

Writing this with radical signs, we have

$$\frac{-4x\pi\sqrt[4]{x^3} + (x^2+\pi)\sqrt{x^2+\pi}}{4(x^2+\pi)\sqrt[4]{x^3}\sqrt{x^2+\pi}}.$$

7. We want to use the trigonometric and chain rules to find the derivative dy/dx of $\pi\cos(3x^3)$. The rule we need to use is $(\cos u)' = -(\sin u)u'$. Thus, $(\pi\cos(3x^3))' = -\pi\sin(3x^3)9x^2$. Given this derivative, we can determine the rate of change at $x = 0$ simply by inserting $x = 0$ into the formula for the derivative. The answer is zero.

8. Given $y = 4 \sin(x^3 + 2x^2 - 1)$, we will use the rule $(\sin u)' = (\cos u)u'$ to find the derivative. Therefore, $y' = 4 \cos(x^3 + 2x^2 - 1) \times (3x^2 + 4x) = 4(3x^2 + 4x)\cos(x^3 + 2x^2 - 1)$.

9. We are given

$$y = \sqrt{1 + \sqrt{1 + \sqrt{1 + x}}}.$$

In order to find the derivative, we'll have to make repeated use of the chain rule. Here's how. First we'll rewrite the formula so that it looks like

$$y = \left\{ 1 + \left[1 + (1 + x)^{\frac{1}{2}} \right]^{\frac{1}{2}} \right\}^{\frac{1}{2}}.$$

The chain rule tells that $\frac{d}{dx} f(g(x)) = f'[g(x)] \times g'(x)$. Here we can let f equal the outer square root function, and g is what's inside:

$$g = 1 + \left[1 + (1 + x)^{\frac{1}{2}} \right]^{\frac{1}{2}}.$$

Therefore, the derivative is

$$y' = \frac{1}{2} \left\{ 1 + \left[1 + (1 + x)^{\frac{1}{2}} \right]^{\frac{1}{2}} \right\}^{-\frac{1}{2}} \times \left\{ 1 + \left[1 + (1 + x)^{\frac{1}{2}} \right]^{\frac{1}{2}} \right\}'.$$

Another way of writing this is

$$y' = \frac{1}{2} \left\{ 1 + \left[1 + (1 + x)^{\frac{1}{2}} \right]^{\frac{1}{2}} \right\}^{-\frac{1}{2}} \times \frac{d}{dx} \left\{ 1 + \left[1 + (1 + x)^{\frac{1}{2}} \right]^{\frac{1}{2}} \right\}.$$

We have to use the chain rule again to compute the derivative that remains on the right of the multiplication symbol:

$$y' = \frac{1}{2} \left\{ 1 + \left[1 + (1 + x)^{\frac{1}{2}} \right]^{\frac{1}{2}} \right\}^{-\frac{1}{2}} \times \frac{1}{2} \left[1 + (1 + x)^{\frac{1}{2}} \right]^{-\frac{1}{2}} \frac{d}{dx} \left[1 + (1 + x)^{\frac{1}{2}} \right].$$

Finally, we have

$$y' = \frac{1}{2}\left\{1+\left[1+(1+x)^{\frac{1}{2}}\right]^{\frac{1}{2}}\right\}^{-\frac{1}{2}} \times \frac{1}{2}\left[1+(1+x)^{\frac{1}{2}}\right]^{-\frac{1}{2}} \times \frac{1}{2}(1+x)^{-\frac{1}{2}},$$

or we can multiply the three ½s to get

$$y' = \frac{1}{8}\left\{1+\left[1+(1+x)^{\frac{1}{2}}\right]^{\frac{1}{2}}\right\}^{-\frac{1}{2}} \times \left[1+(1+x)^{\frac{1}{2}}\right]^{-\frac{1}{2}} \times (1+x)^{-\frac{1}{2}}$$

or

$$y' = \frac{1}{8}\left\{1+\sqrt{1+\sqrt{1+x}}\right\}^{-\frac{1}{2}} \times \left[1+\sqrt{1+x}\right]^{-\frac{1}{2}} \times \frac{1}{\sqrt{1+x}}.$$

See if you can massage the formula further by eliminating the remaining powers of –1/2. Isn't that a delicious dish with which to impress family and friends?

10. Rewrite

$$y = \sqrt{2+\sqrt{1+x}} \text{ as } y = \left(2+(1+x)^{\frac{1}{2}}\right)^{\frac{1}{2}}$$

and make repeated use of the chain rule to find the derivative. Therefore, the derivative is

$$y' = \frac{1}{2}\left(2+(1+x)^{\frac{1}{2}}\right)^{-\frac{1}{2}} \times \frac{1}{2}(1+x)^{-\frac{1}{2}} \times 1,$$

which simplifies to

$$y' = \frac{1}{4\sqrt{1+x}\sqrt{2+\sqrt{1+x}}} = \frac{1}{4\sqrt{(1+x)\left(2+\sqrt{1+x}\right)}}.$$

6. Maxima and Minima

1. We are given $y = x^3$. To find the critical points, we must first find the derivative, which is $3x^2$. The derivative is zero when $x = 0$. This means that the point $(0, 0)$ is a critical point. However, we do not yet know if it is a minimum, maximum, or inflection point. To find out, we take the second derivative, which is $6x$. When x is zero, the second derivative is zero. What do we do? This point is neither a maximum nor a minimum. This solution is simple. The point is an inflection point—the resting point as you climb a mountain. As an extra exercise, draw the curve $y = x^3$ and take a look at the inflection point at $(0, 0)$.

2. Given $y = 3x^3 + 1$, the derivative is $9x^2$. The derivative is zero when $x = 0$, so the point $(0, 1)$ is a critical point. The second derivative, $18x$, will determine whether it is a minimum, maximum, or inflection point. When $x = 0$, the second derivative is also zero. Since this is neither positive nor negative and the sign y'' changes at this point, we have an inflection point. More precisely, we know that this is an inflection point because $y'' > 0$, or concave up, when using values of x immediately to the right of zero, while $y'' < 0$, or concave down, for values of x immediately to the left of zero. Therefore, we have an inflection point. Take a look at the graph of $y = 3x^3 + 1$ and view the inflection point at $(0, 1)$.

3. We want to find the critical points for $f(x) = 4 - 3x + x^2$. First we find the derivative, which is $f'(x) = -3 + 2x$. The derivative is zero only when $x = 3/2$. This means that the only critical point is at $x = 3/2$. To understand the nature of the critical point, we take the second derivative, which is $f''(x) = 2$. At the critical point $x = 3/2$, the second derivative is positive. (In fact, as you can see, the second derivative is always positive.) So, by the second-derivative test, $f(x)$ has a relative minimum at $x = 3/2$. This curve can't have an inflection point, because inflection points are those where the curve changes concavity from concave down to concave up. At these points, $f''(x) = 0$ or $f''(x)$ does not exist, neither of which is the case here.

4. We will need to find the critical points for $f(x) = 5 - 6x + x^2$. The derivative, $f'(x) = -6 + 2x$, is zero only when $x = 3$. Therefore, our critical point is at $(3, -4)$. The second derivative, $f''(x) = 2$, is positive, so this curve has a minimum at $(3, -4)$.

5. We want to find the local maxima, minima, and inflection points of

$$\frac{1}{3}x^3 + \frac{3}{2}x^2 + 2x + 1.$$

First we compute the derivative, which is $y' = x^2 + 3x + 2$. The easiest way to determine where the derivative is zero is to recognize that this factors to $y' = (x + 1)(x + 2)$. Thus, the first derivative is zero at points $x = -1$ and $x = -2$. The second derivative is $y'' = 2x + 3$. At the point $x = -1$, the second derivative is positive, so this point is a local minimum. At the point $x = -2$, the second derivative is negative, so this point is a local maximum.

Does this curve have any potential inflection points? Let us check to see if the second derivative is ever equal to zero. Yes, $y'' = 2x + 3$ is zero when x is $-3/2$. To the left of this point, $f''(x) < 0$, which means the curve is concave down. To the right of this point, $f''(x) > 0$, which means the curve is concave up. Thus, we have an inflection point at $x = -3/2$.

6. We will need to find the critical points for

$$\frac{1}{3}x^3 + \frac{5}{2}x^2 + 6x.$$

The first derivative, $x^2 + 5x + 6$, can also be written as $(x + 2)(x + 3)$. This derivative is zero only when $x = -2$ or $x = -3$. The second derivative is $2x + 5$. When $x = -2$, the second derivative is positive, so this point is a local minimum. At the point when $x = -3$, the second derivative is negative, so this point is a local maximum. An inflection point(s) could exist whenever the second derivative is equal to zero. This happens at $x = -5/2$ or $(-5/2, -4^{7}/_{12})$. To the left of this point, $f''(x) < 0$, which means the curve is concave down. To the right of this point, $f''(x) > 0$, which means the curve is concave up. Therefore, the inflection point is at $x = -5/2$.

7. The second derivative tells us about a curve's concavity. For $x^3 - 9x^2 - 2$, we find $y' = 3x^2 - 18x$, and $y'' = 6x - 18$. The second derivative is less than zero when x is less than 3. This means the curve is concave down for values less than 3. Similarly, the curve is concave up for values of x greater than 3. Note that at $x = 3$, we have an inflection point.

8. For $y = x^3 - 3x - 25$, we find $y' = 3x^2 - 3$ and $y'' = 6x$. The second derivative is less than zero when x is less than zero. Therefore, the curve is concave down for values less than zero. Similarly, the curve is concave up for values greater than zero. An inflection point exists at $x = 0$.

9. The derivative of our function $f(x) = x - kx^{-1}$ is $1 + kx^{-2}$. Because we want $x = -4$ to be a critical number, this means we want the derivative to equal zero at $x = -4$:

$$1 + k\frac{1}{x^2} = 0; 1 + k\frac{1}{16} = 0.$$

And k therefore is -16. Let's look again at the first derivative when we substitute $k = -16$: $f'(x) = 1 - 16x^{-2}$. We can find the second derivative: $f''(x) = 32x^{-3}$. At the location $x = -4$, the first derivative is zero and the second derivative is negative. This means we have a relative maximum at $x = -4$ when k is -16.

10. The derivative of our function

$$f(x) = x - \frac{k}{x}$$

is $1 + kx^{-2}$. Because we want $x = -1$ to be a critical number, the derivative must equal zero at $x = -1$. Therefore, letting $x = -1$, we have

$$1 + \frac{k}{(-1)^2} = 0$$

or $1 + k = 0$. Solving for k, we have $k = -1$. The first derivative, $1 + kx^{-2}$, becomes $f'(x) = 1 - x^{-2}$. At $x = -1$, the second derivative, $f''(x) = 2x^{-3}$, is negative. Thus, we have a relative maximum at $x = -1$ when $k = -1$.

7. Min-Max Pizza Applications

1. The derivative of

$$f(x) = 2x^3 - 4x^2 + 2 \text{ is } f'(x) = 6x^2 - 8x = 2x(3x - 4).$$

Therefore, the critical points are $x = 0$ and $x = 4/3$. However, recall that we want to find the absolute maximum and minimum on the closed interval

[–1, 1], and 4/3 is outside this interval. Thus, for this problem, we don't really care about the behavior of the function at 4/3. So, what we really need to compare are the values of the function at $x = 0$ and at the range's endpoints –1 and 1. Let's put these values in a table to keep track of them:

x	–1	1	0
$f(x)$	–4	0	2

We can see from our table that the absolute minimum is $(-1, -4)$ and the absolute maximum is $(0, 2)$. For curiosity, we can see the second derivative, $f''(x) = 12x - 8$, is negative at $x = 0$, suggesting that this point is a relative, or local, maximum as determined by lessons we learned in chapter 6.

2. The derivative of $f(x) = 8x^3 - 16x^2 + 2$ is $f'(x) = 24x^2 - 32x$. When $f'(x) = 0$, we have the critical points $x = 0$ and $x = 4/3$. The value $x = 4/3$ is outside our interval of $[-1, 1]$. Comparing the values of the function at $x = 0$ and the range's end points –1 and 1, we have

x	–1	1	0
$f(x)$	–22	–6	2

From the table, the absolute minimum is $(-1, -22)$ and the absolute maximum is $(0, 2)$. When $x = 0$, the second derivative, $f''(x) = 48x - 32$, is negative. Therefore, $(0, 2)$ is a relative, or local, maximum.

3. Let x be the horizontal dimension of the pizza and y be the vertical dimension, as illustrated in the top view of the tray in Figure 17. Therefore, the area of the rectangular pizza is xy. The problem stated that the pizza area must be 60 cm², so $xy = 60$. The area of the entire tray is $A = (x + 8)(y + 6)$. We know that $y = 60/x$, so the tray area becomes

$$(x + 8)(60/x + 6) = 6(x + 8)(10/x + 1) = 6(10 + x + 80/x + 8)$$
$$= 6(18 + x + 80/x) = 6(18 + x + 80x^{-1}).$$

We want the pizza to be 60 cm² but to use as little material as possible to make the tray. There is no constraint on x, except, of course, that it can't be a negative number. To find the minimum of the tray's area function, let's take the first and second derivatives of the tray area function:

$$A' = 6(1 - 80/x^2)$$
$$A'' = 960/x^3.$$

To find the critical points of the tray area function, set the first derivative equal to zero:

$$0 = 6(1 - 80/x^2) \text{ or } x = \sqrt{80} \approx 8.944 \text{ cm}.$$

Because the second derivative is positive at $x = 8.944$, this point corresponds to a local minimum for the area function. Note that this is the only critical point, so other values of the tray area function must be larger.

We can also calculate the tray area. Because $xy = 60$, y must be 6.708. Using our formula for tray area, $A = (x + 8)(y + 6)$, we find the area for the entire tray is 215.32 cm².

Out of curiosity, let's randomly try another (x, y) pair to see how much tray material is required. If we let $x = 3$ and $y = 20$, the area of the tray is 11×26, or 286 cm², which, as expected, requires more tray material to be used than our minimum value.

4. Following the method described in problem 3, the area of this new tray would be $A = (x + 8)(y + 6)$. Since $xy = 200$, we have

$$A = (x+8)\left(\frac{200}{x} + 6\right) = 248 + 6x + 1600x^{-1}.$$

Our critical point is found when the derivative, $A' = 6 - 1600x^{-2}$, equals zero, which is at $x = 16.3299$. When $x = 16.3299$, the second derivative, $A'' = 3200x^{-3}$, is positive. Therefore, the critical point is a local minimum. The length x to minimize the amount of tray material used is 16.3299 centimeters.

5. We are told that the base of Luigi's 3-D pizza is a square. Let x be the side of the square base, and let h be the height of the pizza "box." Then $x^2h = 25$. Because the bottom anchovy pizza surface is $4 per square foot, the cost of the bottom is $4x^2$. The cost of the top surface with onions is x^2. The cost of each of the four sides is $2xh$. Therefore, the total cost of the 3-D pizza (not counting any gifts inside) is

$$Cost = 4x^2 + x^2 + 4(2xh) = 5x^2 + 8xh = 5x^2 + 8x(25/x^2) = 5x^2 + 200/x.$$

There is no constraint on the length of side x, except, of course, that it can't be a negative number. As usual, we will calculate the first and second derivatives to determine the existence and nature of critical points:

$$Cost' = 10x - 200/x=2$$
$$Cost'' = 10 + 400/x^3.$$

Where is the first derivative zero? $10x - 200/x^2 = 0$. Multiply the equation by x^2 to get $10x^3 = 200$. Thus, $x^3 = 20$, and x is about 2.71. This means that $x \approx 2.71$ is the only critical point. Because the second derivative is always positive for $x > 0$, $x \approx 2.71$ corresponds to a minimum. Thus, Luigi's 3-D pizza has a base with a 2.71-foot side. The pizza is 3.4 feet tall. We get the height from $x^2h = 25$, or $h = 25/2.71^2 = 3.4$ feet.

6. Following the method described in problem 5, the cost of this new pizza would be $C = 6x^2 + 3x^2 + 16xh$, where $6x^2$ is the cost of the bottom, $3x^2$ is the cost of the top, and $16xh$ is the cost of the sides. The volume of the new pizza is $V = hx^2 = 25$ ft^3, so $h = 25/x^2$. Therefore, our cost equation becomes $C = 9x^2 + 400x^{-1}$. To find the minimum cost, set the derivative, $C' = 18x - 400x^{-2}$, equal to zero. We obtain a critical point at $x = 2.811$. The second derivative, $C'' = 18 + 800x^{-3}$, is positive at $x = 2.811$, so this is a local minimum. If $x = 2.811$, the height of the pizza is 3.164 feet. Therefore, the dimensions that will minimize Luigi's cost are 2.811 feet for the base and 3.164 feet for the height.

7. Luigi's total income every day is $x(9 - 0.004x)$. In other words, this is the income Luigi makes over time as he sells x pizzas. Thus, the profit $P = x(9 - 0.004x) - (4x + 1500) = 5x - 0.004x^2 - 1500$. Let's calculate the first and second derivatives:

$$P' = 5 - 0.008x$$
$$P'' = -0.008.$$

We have a single critical point at $5 - 0.008x = 0$, or $x = 625$. Because the second derivative is negative, this means we have a maximum at $x = 625$. This means that Luigi must sell 625 sushi pizzas each day to maximize his profit.

8. The cost of selling pizzas is $3x + 1000$ dollars and the price of the sushi pizza is $9 - 0.004x$ dollars. Therefore, the revenue would be $R = x(9 - 0.004x)$. Profit is revenue minus cost, so we have $P = 6x - 0.004x^2 - 1000$. To minimize, we need the critical point, which is found when the derivative, $P' = 6 - 0.008x$, equals zero. This is $x = 750$. Since our second derivative, $P'' = -0.008$, is always negative, $x = 750$ will be a maximum. Luigi would need to sell 750 pizzas in order to maximize his profit.

9. We wish to maximize $f(x) = 2x - x^2$ for $x > 0$. To understand the critical points, we take the first and second derivatives:

$$f'(x) = 2 - 2x$$
$$f''(x) = -2.$$

Therefore, the only critical point occurs at $x = 1$. Because the second derivative is negative, $x = 1$ is the maximum. Just for fun, examine other numbers around $x = 1$ to verify they yield lesser differences.

10. The function that describes Fiona's favorite number is $f(x) = x - x^2$. When the derivative, $f'(x) = 1 - 2x$, is equal to zero, a possible number is $x = 1/2$. Since the second derivative, $f''(x) = -2$, is always negative, $x = 1/2$ is a maximum. Therefore, Fiona's favorite number is $1/2$.

8. Exponentials and Logarithms

1. $\left(2e^{\sqrt{x}}\right)' = 2\left(e^{x^{\frac{1}{2}}}\right)' = 2\left(e^{x^{\frac{1}{2}}}\right)\left(\frac{1}{2}x^{-\frac{1}{2}}\right) = \frac{1}{\sqrt{x}}e^{\sqrt{x}}.$

2. $\left(\pi e^{\sqrt{x}}\right)' = \pi \times \frac{1}{2} \times x^{-\frac{1}{2}}e^{\sqrt{x}} = \frac{\pi e^{\sqrt{x}}}{2\sqrt{x}}$

3. $(3^{x^2})' = (\ln 3)(2x)(3^{x^2})$. Here we use the exponential base chain rule $(b^u)' = (\ln b)b^u u'$.

4. $\left(4^{x^3}\right)' = (\ln 4)\left(4^{x^3}\right)(3x^2)$. Again, we use the exponential base chain rule $(b^u)' = (\ln b)b^u u'$.

5. $\left[\ln\!\left(8x^2 + 2\right)\right]' = \dfrac{1}{8x^2 + 2}(16x) = \dfrac{8x}{4x^2 + 1}.$

Here we use the logarithmic chain rule $(\ln u)' = (1/u)u'$.

6. $\left[\ln\!\left(10x^2 - 5\right)\right]' = \dfrac{1}{10x^2 - 5} \times 20x = \dfrac{20x}{10x^2 - 5} = \dfrac{4x}{2x^2 - 1}$

7. $(-e^{\sin x})' = -e^{\sin x}\cos x$

8. $(e^{-2\sin x})' = e^{-2\sin x}(-2\cos x)$

9. $\left[\ln\!\left(e^x + x\right)\right]' = \dfrac{e^x + 1}{e^x + x}$

10. $\left[\pi\ln\!\left(e^x + x^2\right)\right]' = \pi \times \dfrac{1}{e^x + x^2} \times (e^x + 2x) = \dfrac{\pi(e^x + 2x)}{e^x + x^2}$

9. Limits and Continuity

1. We are given

$$\lim_{x \to 6} \frac{x^2 - 36}{x - 6}.$$

The numerator and denominator are both approaching zero as x approaches 6. One way to determine the limit is to recast the numerator:

$$\lim_{x \to 6} \frac{x^2 - 36}{x - 6} = \lim_{x \to 6} \frac{(x+6)(x-6)}{x - 6} = \lim_{x \to 6} (x+6) = 12.$$

2. Given

$$\lim_{x \to 7} \frac{x^2 - 49}{x - 7},$$

the numerator and the denominator are both approaching zero as x approaches 7. The limit is determined as follows:

$$\lim_{x \to 7} \frac{x^2 - 49}{x - 7} = \lim_{x \to 7} \frac{(x+7)(x-7)}{x - 7} = \lim_{x \to 7} (x+7) = 14.$$

3. We are given

$$\lim_{x \to \infty} \frac{5x^2 - 3x + 2}{2x^2 + 7x + 3}.$$

We may divide the numerator and denominator by x^2 to place the fraction in a more manageable form:

$$\lim_{x \to \infty} \frac{5x^2 - 3x + 2}{2x^2 + 7x + 3} = \lim_{x \to \infty} \frac{5 - \frac{3}{x} + \frac{2}{x^2}}{2 + \frac{7}{x} + \frac{3}{x^2}} = \frac{\lim_{x \to \infty} \left(5 - \frac{3}{x} + \frac{2}{x^2}\right)}{\lim_{x \to \infty} \left(2 + \frac{7}{x} + \frac{3}{x^2}\right)}.$$

Because all terms with x and x^2 go to zero as x approaches infinity, the limit is 5/2.

4. Divide the numerator and denominator by x^2 in order to determine the limit:

$$\lim_{x\to\infty}\frac{7x^2-2x+1}{3x^2+5x}=\lim_{x\to\infty}\frac{7-\frac{2}{x}+\frac{1}{x^2}}{3+\frac{5}{x}}=\frac{\lim_{x\to\infty}\left(7-\frac{2}{x}+\frac{1}{x^2}\right)}{\lim_{x\to\infty}\left(3+\frac{5}{x}\right)}=\frac{7}{3}.$$

5. We are given

$$\lim_{x\to0}\frac{2x}{\sqrt{x+1}-1}.$$

Let's multiply the numerator and denominator by $\sqrt{x+1}+1$ to simplify the fraction:

$$\lim_{x\to0}\frac{2x}{\sqrt{x+1}-1}=\lim_{x\to0}\frac{2x}{\sqrt{x+1}-1}\times\frac{\sqrt{x+1}+1}{\sqrt{x+1}+1}$$

$$=\lim_{x\to0}\frac{2x\left(\sqrt{x+1}+1\right)}{(x+1)-1}=\frac{\lim_{x\to0}\left[2\left(\sqrt{x+1}+1\right)\right]}{\lim_{x\to0}(1)}=4.$$

6. Simplify the fraction by multiplying by $\sqrt{x+1}+1$ in order to determine the limit:

$$\lim_{x\to0}\frac{\pi x}{\sqrt{x+1}-1}=\lim_{x\to0}\frac{\pi x}{\sqrt{x+1}-1}\times\frac{\sqrt{x+1}+1}{\sqrt{x+1}+1}=\lim_{x\to0}\frac{\pi x\sqrt{x+1}+\pi x}{x+1-1}$$

$$=\lim_{x\to0}\frac{\pi x\sqrt{x+1}+\pi x}{x}=\frac{\lim_{x\to0}\left(\pi\sqrt{x+1}+\pi\right)}{\lim_{x\to0}(1)}=2\pi.$$

7. We are given

$$\lim_{x\to0}\frac{1-e^x}{2x}.$$

It is appropriate to use L'Hôpital's rule because both the numerator and denominator are zero when $x = 0$:

$$\lim_{x\to0}\frac{1-e^x}{2x}=\lim_{x\to0}\frac{-e^x}{2}=-\frac{1}{2}.$$

8. Use L'Hôpital's rule to find the limit, since we have 0/0 when $x = 0$:

$$\lim_{x\to 0}\frac{1-e^x}{3x} = \lim_{x\to 0}\frac{-e^x}{3} = -\frac{1}{3}.$$

9. We are given L'Hôpital's rule to find the limit

$$\lim_{x\to\infty}\frac{2\ln x}{x}.$$

It is appropriate to use L'Hôpital's rule because both the numerator and denominator are infinite when $x = \infty$:

$$\lim_{x\to +\infty}\frac{2\ln x}{x} = \lim_{x\to +\infty}\frac{2\times 1/x}{1} = 0.$$

10. Use L'Hôpital's rule to find the limit, since we have ∞/∞ when $x = 0$.

$$\lim_{x\to\infty}\frac{\ln x}{2x} = \lim_{x\to\infty}\frac{1/x}{2} = \frac{0}{2} = 0.$$

10. Related Rates

1. Let a be the altitude of the meatball, and let w be the distance of the meatball from the ground along its flight path (Figure 39). Then

$$\frac{a}{w} = \sin 45° = \frac{\sqrt{2}}{2}$$

or

$$2a = \sqrt{2}w.$$

We may take the derivative of both sides with respect to time:

$$2\frac{d}{dt}a = \sqrt{2}\frac{d}{dt}w.$$

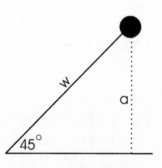

FIGURE 39. Schematic diagram indicating w and a.

Since we know that the meatball is flying at 200 kilometers per hour along the flight path, we have

$$2\frac{d}{dt}a = \sqrt{2} \times 200$$

or

$$\frac{d}{dt}a = a' = \sqrt{2} \times 100.$$

This means that the altitude is changing at a rate of about 141.4 kilometers per hour.

2. If the meatball was traveling at a 60° angle with respect to the ground, we would have

$$\sin 60 = \frac{a}{w}.$$

This simplifies to $\sqrt{3}w = 2a$. Taking the derivative of both sides with respect to time, we have

$$\sqrt{3}\frac{dw}{dt} = 2\frac{da}{dt}.$$

Since the meatball is flying at 200 kilometers per hour, we have

$$200 \times \sqrt{3} = 2\frac{da}{dt} \quad \text{or} \quad \frac{da}{dt} = 100\sqrt{3}.$$

Therefore, the change in altitude with respect to time is $100\sqrt{3}$ or 173.2 kilometers per hour.

3. Let V be the volume of tomato juice at time t, and let h be the depth of the tomato juice in the huge can. Therefore, $V = \pi 100^2 h$. Let's take the derivative of both sides of this equation:

$$\frac{d}{dt}V = 10,000\pi\frac{d}{dt}h.$$

(Recall that

$$\frac{d}{dt}V$$

is just another way of writing V', where the derivative is calculated with respect to time.) We know that the can is being filled with tomato sauce at the rate of 10,000 cubic feet per minute. This means that

$$\frac{d}{dt}V = 10,000 \text{ cubic feet per minute.}$$

Therefore, $10,000 = 10,000\pi h'$ or $10,000/(10,000\pi) = h'$. Thus, the depth of the tomato sauce is increasing at the rate of $1/\pi$ feet per minute.

4. If the can had a radius of 20 feet, the volume of the can would be $V = \pi(20)^2 h$. The derivative would be

$$\frac{dV}{dt} = 400\pi\frac{dh}{dt}.$$

Since the rate

$$\frac{dV}{dt}$$

at which the can is filling is 10,000 cubic feet per minute, we have

$$\frac{dh}{dt} = \frac{10,000}{400\pi} = \frac{25}{\pi}.$$

The depth of the tomato sauce is increasing at a rate of $25/\pi$ feet per minute.

5. Of course, it is unrealistic to be thinking about a 100-mile-long track resting against a brick wall, because, for one thing, it would be difficult to make a rigid structure that long, but Luigi likes dreaming up odd problems. Let y be the distance of the top of the track from the ground. Let x be the distance of the bottom of the track B from the brick wall. For the triangle formed by the track, wall, and ground, we know $x^2 + y^2 = 100^2$. Let's try to differentiate each term of this equation with respect to time:

$$\frac{d}{dt}x^2 + \frac{d}{dt}y^2 = \frac{d}{dt}100^2$$

or

$$2x\frac{d}{dt}x + 2y\frac{d}{dt}y = 0.$$

Once we divide all terms by 2, we can rewrite this as $xx' + yy' = 0$. We know the track is sliding down the wall at a remarkable speed of 1 mile per second; thus, $y' = -1$. The derivative is negative because y is continually decreasing. Because we are interested in solving the problem when B is 10 miles away from the base of the wall, $x = 10$. Using the Pythagorean theorem, $x^2 + y^2 = 100^2$, we have $y^2 = 10,000 - 100$, which means that y is roughly equal to 99.498. Thus, we have $10x' + (-1)(99.498) = 0$ and $10x' = (99.498)$ or $x' = 9.9498$ miles/second. Notice how much faster point B moves than the top of the ladder, which is moving only at 1 mile per second. Luigi had better get out of the way fast!

Because point B moves faster to the right than the top of the ladder moves down, does this mean that if we moved the top of the track down at speeds close to the speed of light that point B would be moving faster than the speed of light? Physicists tells us it should be impossible to move an object faster than the speed of light, but what prevents this in the sliding track scenario?

6. When the track is 10 miles long, we have the equation $x^2 + y^2 = 100$. Taking the derivative of this equation results in $2xx' + 2yy' = 0$. Since the track is sliding at a rate of 1 mile per second, we have $y' = -1$. Also, since we want the speed at the bottom of the track when B is 5 miles away from the wall, we have $x = 5$. Therefore, solving the equation, $5^2 + y^2 = 100$ for y gives $y = \sqrt{75} = 5\sqrt{3}$. Evaluate $2xx' + 2yy' = 0$ for x', letting $x = 5$, $y' = -1$, and $y = 5\sqrt{3}$. The speed at the bottom of the track when B is 5 miles away from the wall is $\sqrt{3}$ or approximately 1.732 miles per second.

7. The area of the pizza is $A = \pi r^2$. Let's take the derivative of both sides of the equation with respect to time: $A' = (\pi r^2)' = 2\pi r \times r'$. We know that $r' = 3$. This means $A' = 2\pi r \times 3$. When r is 200, the rate of increase in the pizza's area is $A' = 2\pi 200 \times 3 \approx 3{,}769.9$ square inches per minute. Big Tony was wise to run from the restaurant. I wonder how long it would take the growing pizza to envelop the earth?

8. The area of the pizza is $A = \pi r^2$ and the change in area is the derivative of the area formula,

$$A' = 2\pi r \frac{dt}{dt}.$$

When the radius of the pizza is 10 inches and growing at a rate of half an inch per minute, we have $A' = 2\pi(10)(0.5) = 10\pi = 31.416$. Therefore, the area of the pizza is growing at a rate of 31.416 in²/min.

9. Let's first take the derivative of the ant's path with respect to time:

$$y = x^2 + 3x - 1$$

$$\frac{d}{dt}y = 2x \times \frac{d}{dt}x + 3 \times \frac{d}{dt}x,$$

which we can rewrite more simply as $y' = 2x \times x' + 3 \times x'$, and we remember that the derivatives are with respect to time. Thus, $y' = x'(2x + 3)$. Our problem is to determine where the x- and y-coordinates of the ant are changing at the same rate—that is, we want to know when $y' = x'$. We can make a substitution in $y' = x'(2x + 3)$, so $y' = y'(2x + 3)$, and, therefore, $2x + 3 = 1$, so $x = -1$. Substituting this value into our original equation $y = x^2 + 3x - 1$, we find that $y = -3$. So, the x- and y-coordinates of the ant are changing at the same rate at $(-1, -3)$. Note that if the ant is sitting still, $y' = x'$ no matter where the ant is on the curve.

10. Given the ant's path of $y = x^2 + 4x - 3$, take the derivative of the ant's path with respect to time. This is $y' = 2xx' + 4x'$, which can be written as $y' = x'(2x + 4)$. When the x- and y-coordinates are changing at the same rate, we have $y' = x'$. Therefore, $x' = x'(2x + 4)$ or $x = -3/2$. At $x = -3/2$, $y = -27/4$, so the x- and y-coordinates are changing at the same rate at the point $(-3/2, -27/4)$.

11. Integration

Antiderivatives (Indefinite Integrals)

1. We are given

$$\int \left(7 + \frac{1}{\sqrt{x}} \right) dx.$$

First let's put this in a form that is more easy to handle:

$$\int (7 + x^{-\frac{1}{2}}) dx.$$

We can now take the integral $7x + 2x^{\frac{1}{2}} + C = 7x + 2\sqrt{x} + C.$

2. Rewrite the square root as a rational fraction in order to find the integral

$$\int \left(\pi + \frac{2}{\sqrt{x}} \right) dx = \int \left(\pi + 2x^{-\frac{1}{2}} \right) dx = \pi x + 4x^{\frac{1}{2}} + C = \pi x + 4\sqrt{x} + C.$$

3. We are given

$$\int (2\sin x - 4\cos x + 2x^3) dx.$$

We can use our rules of integration for sine and cosine to solve the first two terms:

$$\int (2\sin x - 4\cos x + 2x^3) dx = 2(-\cos x) - 4\sin x + 2\left(\frac{x^4}{4} \right) + C$$

$$= -2\cos x - 4\sin x + \frac{x^4}{2} + C.$$

4. Use the rules of integration for sine and cosine to solve the first two terms:

$$\int (6\cos x - 4\sin x - 3x^4)dx = 6(\sin x) - 4(-\cos x) - \frac{3}{5}x^5 + C$$

$$= 6\sin x + 4\cos x - \frac{3}{5}x^5 + C.$$

5. We are given

$$\int \sqrt{5x-2}\, dx.$$

We can't use our simple equation such as

$$\int x^r dx = \frac{x^{r+1}}{r+1} + C$$

directly on this because we don't have an integrand that is actually in the form of x^r. To solve this problem, we first have to make substitutions. If we let $u = 5x - 2$, then $du = 5dx$. Thus, we can simplify the integral and put it in the form

$$\int \sqrt{5x-2}\, dx = \int \sqrt{u}\,\frac{1}{5}du.$$

This is the same as

$$\frac{1}{5}\int \sqrt{u}\, du = \frac{1}{5}\int u^{\frac{1}{2}}du = \frac{1}{5}\left(\frac{2}{3}\right)u^{\frac{3}{2}} + C$$

$$= \left(\frac{2}{15}\right)(5x-2)^{\frac{3}{2}} + C = \left(\frac{2}{15}\right)\left(\sqrt{5x-2}\right)^3 + C.$$

6. We are given

$$\int \sqrt{2x+4}\,dx,$$

which equals

$$\int (2x+4)^{\frac{1}{2}}dx.$$

Let $u = 2x + 4$ so that $du = 2dx$. Simplify the integral by putting it in the form

$$\int (2x+4)^{\frac{1}{2}}dx = \int \frac{1}{2}u^{\frac{1}{2}}du.$$

This is the same as

$$\frac{1}{2}\int u^{\frac{1}{2}}du = \frac{1}{2}\times\frac{2}{3}u^{\frac{3}{2}}+C = \frac{1}{3}u^{\frac{3}{2}}+C = \frac{1}{3}(2x+4)^{\frac{3}{2}}+C$$

$$= \frac{1}{3}\left(\sqrt{2x+4}\right)^{3}+C.$$

Definite Integrals

7. We are given

$$\int_{1}^{3}(6x^2 + x + 1)dx.$$

We know that

$$\int (6x^2 + x + 1)dx = 2x^3 + \frac{x^2}{2} + x + C.$$

This means that

$$\int_{1}^{3}(6x^2 + x + 1)dx = \left[2x^3 + \frac{x^2}{2} + x\right]_{1}^{3}.$$

Notice how we will omit the constant because it will subtract out of the computation once we evaluate the terms from 1 to 3:

$$\left[2x^3 + \frac{x^2}{2} + x\right]_{1}^{3} = \left(2\times3^3 + \frac{3^2}{2} + 3\right) - \left[2\times1^3 + \frac{1^2}{2} + 1\right] = 61.5 - 3.5 = 58.$$

8. We are given

$$\int_{1}^{3}(9x^2 + 2x - 1)dx.$$

We solve as follows:

$$\left[3x^3 + x^2 - x\right]_1^3 = (3 \times 3^3 + 3^2 - 3) - (3 \times 1^3 + 1^2 - 1) = 87 - 3 = 84.$$

9. We are given

$$\int_0^1 \frac{x}{(3x^2 - 1)^3}\, dx.$$

Here is a problem like problem 5 in which we have to make a substitution. Let $u = 3x^2 - 1$, which means that $du = 6x\,dx$. We can simplify our original expression:

$$\int_0^1 \frac{x}{(3x^2 - 1)^3}\, dx = \int_{-1}^2 \frac{1/6}{u^3}\, du.$$

Notice how we changed the integral's range from (0 to 1) to (–1 to 2) because we switched variables from x to u. Now we can simplify further:

$$\frac{1}{6}\int_{-1}^2 u^{-3}\, du = \left[-\frac{1}{12} u^{-2}\right]_{-1}^2 = \left[-\frac{1}{12u^2}\right]_{-1}^2 = -\frac{1}{12}\left(\frac{1}{4} - 1\right) = \frac{1}{16}.$$

10. We are given

$$\int_3^4 \frac{x}{(2x^2 + 4)^3}\, dx.$$

Let $u = 2x^2 + 4$ so that $du = 4x\,dx$ or $\frac{1}{4}du = x\,dx$. When we substitute, we must change the limits from (3 to 4) to (22 to 36). Therefore,

$$\int_{22}^{36} \frac{1}{4} u^{-3}\, du = \left[\frac{1}{4} \times \frac{-1}{2} u^{-2}\right]_{22}^{36} = \left[\frac{-1}{8u^2}\right]_{22}^{36}$$

$$= \frac{-1}{8 \times 36^2} - \frac{-1}{8 \times 22^2} = \frac{-1}{10{,}368} + \frac{1}{3{,}872} = \frac{203}{1{,}254{,}528}.$$

12. Logarithmic Differentiation, Integration by Parts, Trigonometric Substitution, and Partial Fractions

1. We are asked to use logarithmic differentiation to find the derivative of $y = x^3\sqrt{2+x^2}$. We first take the logarithm of both sides of the equation, remembering that $\ln ab = \ln a + \ln b$: $\ln y = \ln x^3 + \ln(2 + x^2)^{\frac{1}{2}}$. This is the same as

$$\ln y = 3\ln x + \frac{1}{2}\ln(2+x^2).$$

Now we use implicit differentiation on both sides of the equation, remembering the rule $(\ln u)' = (1/u)u'$:

$$\frac{1}{y}y' = 3\times\frac{1}{x}+\frac{1}{2}\times\frac{1}{2+x^2}\times 2x$$

or

$$\frac{1}{y}y' = \frac{3}{x}+\frac{x}{2+x^2} = \frac{6+3x^2+x^2}{x(2+x^2)} = \frac{6+4x^2}{x(2+x^2)} = \frac{2(3+2x^2)}{x(2+x^2)}.$$

So, we know

$$y' = \frac{2(3+2x^2)}{x(2+x^2)}\times y = \frac{2(3+2x^2)}{x(2+x^2)}\times x^3(2+x^2)^{\frac{1}{2}} = \frac{2x^2(3+2x^2)}{\sqrt{(2+x^2)}}.$$

2. Given $y = x^3\sqrt{20-5x^2}$, take the log of both sides to obtain

$$\ln y = \ln x^3 + \ln(20-5x^2)^{\frac{1}{2}} = 3\ln x + \frac{1}{2}\ln(20-5x^2).$$

Differentiate both sides, remembering that $\ln u = u'/u$. This gives

$$\frac{y'}{y} = 3\times\frac{1}{x}+\frac{1}{2}\times\frac{-10x}{20-5x^2} = \frac{3}{x}-\frac{x}{4-x^2}.$$

Solving for y', we have

$$y' = \left(x^3\sqrt{20 - 5x^2}\right)\left(\frac{12 - 4x^2}{4x - x^3}\right).$$

3. We are asked to use integration by parts to find

$$\int xe^{4x}\,dx.$$

To do integration by parts, recall the formula

$$\int u\,dv = uv - \int v\,du.$$

Here we can let $u = x$ and $dv = e^{4x}dx$. This means that $du = dx$ and

$$v = \frac{1}{4}e^{4x}.$$

Let's apply the integration by parts formula, remembering that the integral of

$$e^{ax} \text{ is } \frac{1}{a}e^{ax} + C:$$

$$\int xe^{4x}\,dx = \frac{1}{4}xe^{4x} - \int \frac{1}{4}e^{4x}\,dx = \frac{1}{4}xe^{x} - \frac{1}{4}\times\frac{1}{4}e^{4x} + C$$

$$= \frac{1}{16}e^{4x}(4x - 1) + C.$$

4. Given

$$\int xe^{5x}\,dx,$$

use integration by parts to integrate. Let $u = x$ so that $du = dx$; $dv = e^{5x}\,dx$ so that $v = (1/5)e^{5x}$. We now have the following:

$$\int xe^{5x}dx = \frac{1}{5}xe^{5x} - \int \frac{1}{5}e^{5x}dx = \frac{1}{5}xe^{5x} - \frac{1}{25}e^{5x} + C = \frac{1}{25}e^{5x}(5x-1) + C.$$

5. We are asked to use the method of partial fractions to find

$$\int \frac{dx}{x^2 - 16}.$$

One way to break this into two easier-to-handle fractions is to use the following approach:

$$\frac{1}{x^2 - 16} = \frac{1}{(x-4)(x+4)} = \frac{A}{x-4} + \frac{B}{x+4}.$$

If we multiply by $(x - 4)(x + 4)$, we can get rid of the denominators: $1 = A(x + 4) + B(x - 4)$. If we let $x = 4$, then we have $1 - A(4 + 4) + 0$ or $A = 1/8$. If we let $x = -4$, then $1 = -8B$ and $B = -1/8$. This means that

$$\frac{1}{x^2 - 16} = \frac{1}{8} \times \frac{1}{x-4} - \frac{1}{8} \times \frac{1}{x+4}.$$

Now let's look at our original integral that we wish to solve and recall our integral rule:

$$\int u^{-1}du - \ln|u| + C.$$

This means that

$$\int \frac{dx}{x^2 - 16} = \frac{1}{8}\ln|x-4| - \frac{1}{8}\ln|x+4| + C.$$

6. Given

$$\int \frac{dx}{x^2 - 25},$$

use the method of partial fractions to integrate. We have

$$\frac{1}{(x+5)(x-5)} = \frac{A}{x+5} + \frac{B}{x-5}.$$

This becomes $1 = A(x-5) + B(x+5)$. If $x = 5$, then $B = 1/10$. If $x = -5$, then $A = -1/10$. Therefore, we have the following:

$$\int \frac{dx}{x^2 - 25} = \int \frac{-1}{10} \times \frac{1}{x+5} dx + \int \frac{1}{10} \times \frac{1}{x-5} dx$$

$$= \frac{-1}{10} \ln|x+5| + \frac{1}{10} \ln|x-5| + C.$$

7. We are asked to use logarithmic differentiation to calculate the derivative of $f(x) = (x^6+666)(x^7+777)x^8$. Let's start by taking the natural logarithm of both sides of the equation: $\ln f(x) = \ln[(x^6 + 666)(x^7 + 777)x^8]$. Because we know that $\ln(ab) = \ln(a) + \ln(b)$, our big product becomes a sum: $\ln f(x) = \ln(x^6+666) + \ln(x^7+777) + \ln(x^8)$. Next we differentiate both sides, recalling the rule $(\ln u)' = u'/u$. We can rewrite our equation as

$$\frac{f'(x)}{f(x)} = \frac{6x^5}{x^6 + 666} + \frac{7x^6}{x^7 + 777} + \frac{8x^7}{x^8}.$$

Our mission is to find $f'(x)$, so we must multiply both sides of the equation by $f(x)$:

$$f'(x) = f(x)\left(\frac{6x^5}{x^6 + 666} + \frac{7x^6}{x^7 + 777} + \frac{8x^7}{x^8}\right)$$

or

$$f'(x) = (x^6 + 666)(x^7 + 777)x^8\left(\frac{6x^5}{x^6 + 666} + \frac{7x^6}{x^7 + 777} + \frac{8x^7}{x^8}\right).$$

8. Given $f(x) = (x^{20} - \pi)(x^7 - 777)x^3$, use logarithmic differentiation to calculate the derivative. Taking the log of both sides results in $\ln f(x) = \ln(x^{20} - \pi) + \ln(x^7 - 777) + \ln x^3$. Differentiate both sides, remembering that $\ln u = u'/u$. This gives

$$\frac{f'(x)}{f(x)} = \frac{20x^{19}}{x^{20}-\pi} + \frac{7x^6}{x^7-777} + \frac{3}{x}.$$

Solving for $f'(x)$ gives us our answer:

$$f'(x) = (x^{20}-\pi)(x^7-777)x^3 \left[\frac{20x^{19}}{x^{20}-\pi} + \frac{7x^6}{x^7-777} + \frac{3}{x}\right].$$

9. We are asked to solve the following integral using integration by parts:

$$\int \left(\frac{\ln x}{\pi^2 x^2} + \frac{2x^3}{\pi^2}\right) dx.$$

First we can break this integral into two terms:

$$\int \left(\frac{\ln x}{\pi^2 x^2} + \frac{2x^3}{\pi^2}\right) dx = \frac{1}{\pi^2}\left(\int \frac{\ln x}{x^2} dx + \int 2x^3 dx\right).$$

Let's save the $1/\pi^2$ as a multiplier for our result because it's just a constant. The second term is

$$2\frac{x^4}{4} + C = \frac{x^4}{2} + C.$$

Let's also save this for later. To integrate the first term of the formula, we need to do integration by parts and recall the formula

$$\int u\, dv = uv - \int v\, du.$$

If $u = \ln x$, then $du = (1/x)dx$, and $dv = (1/x^2)dx$ and $v = -1/x$. Then we have

$$\underset{u}{\int \frac{\ln x}{x^2}} \underset{dv}{dx} = \int (\ln x) \times \left(\frac{1}{x^2} dx\right) = \ln x \times \left(-\frac{1}{x}\right) - \int \left(-\frac{1}{x}\right)\left(\frac{1}{x} dx\right)$$

$$= \frac{-\ln x}{x} + \int \frac{x}{x^2} dx = \frac{-\ln x}{x} - \frac{1}{x} + C = -\frac{1}{x}(\ln x + 1) + C.$$

Let's not forget to add the value for the integral of the second term to get our final answer. We also can bring back our $1/\pi^2$ multiplier:

$$-\frac{1}{x\pi^2}(\ln x + 1) + \frac{x^4}{2\pi^2} + C.$$

10. Given

$$\int\left(\frac{e^3 \ln x}{x^2} + 25e^3 x^3\right)dx,$$

use integration by parts to integrate. Let $u = \ln x$ so that

$$du = \frac{1}{x}dx; \; dv = \frac{1}{x^2}dx \text{ so that } v = \frac{-1}{x}.$$

We now have the following:

$$\int\left(\frac{e^3 \ln x}{x^2} + 25e^3 x^3\right)dx = \int\frac{e^3 \ln x}{x^2}dx + \int 25e^3 x^3 dx$$

$$= e^3\left[\frac{-\ln x}{x} - \int\frac{-1}{x^2}dx\right] + 25e^3\int x^3 dx = e^3\left[\frac{-\ln x}{x} + \left(-1x^{-1}\right) + C\right]$$

$$+ 25e^3 \times \frac{1}{4}x^4 + C = -e^3\left[\frac{\ln x + 1}{x}\right] + \frac{25}{4}e^3 x^4 + C.$$

13. Exponential Growth and Decay

1. Let $N(t)$ be the number of fungal spores as a function of time. In this chapter, we learned that $N(t) = N_0 e^{kt}$ for exponential growth. For simplicity, we can assign $y = N(t)$, and just remember that y is a function of time. Thus, the equation becomes $y = y_0 e^{kt}$. We know that the initial number of spores has doubled in 6 hours. Thus, $2y_0 = y_0 e^{k6}$ or $2 = e^{6k}$. Let's take the natural logarithm of both sides of this equation to get $\ln 2 = 6k$. Thus, we now know that $k = (\ln 2)/6$.

Our mission is to compare the number of spores present in 12 hours to the number of initial spores. When $t = 12$, we know that $y = y_0 e^{12k}$.

Because we know the value for k, we have $y = y_0 e^{2\ln 2}$. Using fundamental rules of exponents, such as $a^{pq} = (a^p)^q$, we see this equation becomes $y = y_0(e^{\ln 2})^2 = y_0 2^2 = 4y_0$. After 12 hours, we have 4 times the initial number of spores.

2. If the initial number of spores has tripled in 6 hours, our general equation would be $3N_0 = N_0 e^{6k}$. Solving for k by taking the log of both sides, we have

$$k = \frac{\ln 3}{6}.$$

Therefore,

$$N(12) = N_0 e^{t \times \frac{\ln 3}{6}} = N_0 e^{2\ln 3} = N_0 3^2 = 9N_0.$$

3. Let y be the number of olives present at time t. We can use our handy formula from problem 1, namely $y = y_0 e^{kt}$. Because 300 olives "decay" to 50 olives in 1 hour, we have $50 = 300e^k$ or $e^k = 1/6$. Let's take the log of both sides, which yields a negative value of k. Negative values for k are customary for decay problems. Here is what we find: $\ln e^k = \ln 1/6$, $k = \ln 1/6$, so $k = -\ln 6$. (Recall the identity $\ln(a/b) = \ln(a) - \ln(b)$.) We want to know how many olives will be in the jar in 2 hours. Thus, when $t = 2$, we have $y = 300e^{2k} = 300(e^k)^2 = 300(1/6)^2$, which is about 8.3 olives.

4. If 400 olives becomes 300 olives in 1 hour, we have the following equation: $300 = 400e^{1k}$. Solving for k by taking the log of both sides, we have $k = \ln(3/4)$. Therefore, after 2 hours,

$$N(2) = 400e^{2\ln\frac{3}{4}} = 400\left(\frac{3}{4}\right)^2 = 225$$

olives that will remain.

5. Let y be the Italian population in millions in year t, with $t = 0$ in 1999. Then $y = 57e^{0.01t}$. In the year 2029, when $t = 30$, we have $y = 57e^{(0.01)(30)} = 76.94$ million people. (If you don't have a calculator with exponential functions, you can leave your answer in the form $57e^{0.3}$.)

6. For the year 2040, which is 41 years after the initial population recording of 57 million and a growth constant of $k = 0.001$, the population will be $N(41) = 57e^{0.001(41)} \approx 59.39$ million people.

7. We know that $y = y_0 e^{kt}$. We are interested in the half-life point (at time T) at which the number of calzones is half the number originally stored in the warehouse, or $(1/2)y_0 = y_0 e^{kT}$. Let's take the log of both sides to yield $\ln (1/2) = kt$. Thus, the relationship between k and t at the half-life point is $kt = -\ln 2$. (Recall the identity $\ln(a/b) = \ln(a) - \ln(b)$.)

8. The general relationship between k and T at the "third-life" point is found as follows: $(1/3)y_o = y_0 e^{kT}$, which simplifies to $\ln (1/3) = kT$. Therefore, the general relationship between k and T at the "third-life" point is $kT = -\ln 3$.

9. Let y be the number of calzones t days after the rats entered the warehouse We are told the half-life is 100 days. Then $y = y_0 e^{kt}$, where $100k = -\ln 2$. (We know this relationship between k and t from the answer to problem 7.) This means that $k = -\ln(2)/100 = -0.0069$.

When Luigi enters the warehouse at time t, $y = (0.1)y_0$. Thus, $(0.1)y_0 = y_0 e^{kt}$ or $(0.1) = e^{kt}$ or $kt = \ln(1/10)$. We know that $k = -0.0069$. This means that $kt = -\ln 10$ or $t = -(\ln 10)/-0.0069 = 333.7$. Thus, the aliens entered the warehouse about 333.7 days before Luigi discovered their existence. If you solved this challenging problem, you are a great detective!

10. The half-life of the calzones is 200 days and only 5 percent of the original calzones remain. Then utilizing the relationship between k and T, $kT = -\ln 2$, we have $200k = -\ln 2$ so that $k = -\ln 2/200$. Since only 5 percent or 1/20 of the calzones remain, we have

$$\frac{1}{20} = e^{\frac{-\ln 2}{200} t}.$$

Taking the log of both sides, we can solve for t:

$$\ln\left(\frac{1}{20}\right) = \frac{-\ln 2}{200} t$$

or the rats entered the warehouse at

$$t = \frac{200 \ln 20}{\ln 2} = 864.3856$$

days before Luigi entered.

14. Calculus and Computers

Regarding Big Tony's question:

> IF THIS PIZZA WEIGHS
> 10 POUNDS PLUS HALF ITS
> OWN WEIGHT, HOW MUCH
> DOES IT WEIGH?

No, the pizza does not weigh 15 pounds, as some people have suggested to Luigi. And you don't need calculus to solve this. In fact, it's sometimes useful to be able to identify when calculus is needed for solving problems and when it is not.

If you are an educator, one good way to have students work on this problem is to visualize a balance scale. The pizza is on the left side. On the right side is a 10-pound weight and half a pizza. The scale is perfectly balanced. Stop and draw the scale. Now look at the right side of your balance. Notice that the 10-pound weight is, in essence, taking the place of half the pizza, which means another 10-pound weight could take the place of the pizza half. By looking at the drawing, you can see that the pizza weighs 20 pounds. Can you figure this out with algebra? For example, a relevant equation might be $x = 10 + \frac{1}{2}x$. Try showing your friends the value of visualization in problem solving. There's nothing quite like drawing a diagram to illustrate a problem before you attempt to solve it.

17. Luigi's Mind-Boggling Workout Routine

1. The volume of the can in the shape of a cylinder is $V = \pi r^2 h$, or, for our case, $100 = \pi r^2 h$. The surface area of a cylinder is the area of the two bases (each with area πr^2 inches^2) plus the area of the side surface ($2\pi rh$): $A = 2\pi r^2 + 2\pi rh$. Notice how the area depends on two variables, r and h. However, we can use the volume formula to eliminate h:

$$A(r) = 2\pi r^2 + 2\pi r \left(\frac{100}{\pi r^2} \right) = 2\pi r^2 + 200r^{-1}.$$

We also want to minimize the surface area A because this will minimize the material used in manufacturing the can. We are interested in exploring only those values of r greater than zero. Let's find the number and nature of critical points by taking the first derivative:

$$A'(r) = 4\pi r - 200r^{-2} = r^{-2}[4\pi r^3 - 200].$$

The critical point is at

$$r^{-2}[4\pi r^3 - 200] = 0$$

or

$$[4\pi r^3 - 200] = 0$$

or

$$r = \sqrt[3]{\frac{200}{4\pi}} \approx 2.5.$$

To find out if $r = 2.5$ is a maximum or minimum, we determine the second derivative, which is $A''(r) = 4\pi + 400r^{-3}$. Because the second derivative is positive at $r = 2.5$, we have arrived at a minimum in the area function. Thus, Luigi's tomato sauce can should have a radius of about 2.5 inches. Can you determine how many square inches of paint Luigi needs to paint this "optimum" can?

2. If the volume of the can is to be 1,000 cubic inches, then the height of the can would be expressed as

$$h = \frac{1000}{\pi r^2}.$$

Substituting the value of h into the surface area of a can formula, we have

$$A = 2\pi r^2 + 2\pi r \left(\frac{1000}{\pi r^2} \right)$$

or

$$A = 2\pi r^2 + 2000r^{-1}.$$

Find the critical point by setting the derivative of A equal to zero: $A' = 4\pi r - 2000r^{-2} = 4\pi r^3 - 2000 = 0$. Solving for r, we have

$$r = \sqrt[3]{\frac{2000}{4\pi}} = 5.419.$$

Since the second derivative, $A'' = 4\pi + 4000r^{-3}$, is positive at $x = 5.419$, this is the minimum. Therefore, the radius of the can is 5.419 inches.

3. Hello. The relevant equation is $P_1 + 10 = 5P_1 - 2$. The answer is 3. Notice that P_2 was not even needed. I just love this question. Few people solve it very quickly. Notice that you don't need any calculus for this. Make sure you know when calculus is needed to solve problems and when it is not needed. Also, always check for and exclude any extra information that is not really needed to solve a problem.

4. Our equation would be $O + 14 = 5O - 2$. Solving for O, we find that Luigi has 4 olives.

5. We are asked to find

$$\lim_{x \to \infty} \frac{15x^3 - 2x + 10}{5x^3 + \pi}.$$

At first glance, both the numerator and denominator seem to be approaching infinity. One way to solve this problem is to divide both the numerator and denominator by the highest power of x in the denominator—that is, x^3:

$$\lim_{x \to \infty} \frac{15x^3 - 2x + 10}{5x^3 + \pi} = \frac{\lim\limits_{x \to \infty} 15 - \lim\limits_{x \to \infty}(2/x^2) + \lim\limits_{x \to \infty}(10/x^3)}{\lim\limits_{x \to \infty} 5 + \lim\limits_{x \to \infty}(\pi/x^3)}$$

$$= \frac{15 - 0 + 0}{5 + 0} = \frac{15}{5} = 3.$$

6. If

$$\beta = \frac{d}{dx}\left(\frac{1}{\pi}x + e^2\right),$$

then

$$\beta = \frac{1}{\pi}.$$

To determine the limit, divide all terms by x^4, then evaluate:

$$\lim_{x \to \infty} \frac{20x^4 - 25x + \left(\frac{1}{\pi}\right)}{5x^4 + \pi^2} = \frac{\lim_{x \to \infty}\left(20 - \frac{25}{x^3} + \frac{1/\pi}{x^4}\right)}{\lim_{x \to \pi}\left(5 + \frac{\pi^2}{x^4}\right)} = \frac{20 - 0 + 0}{5 + 0} = 4.$$

7. We are asked to find the find the derivative of

$$\frac{1}{(\pi x^3 + 3x^2 + 20x + 5)^4}.$$

One way to solve this is to use the chain rule:

$$\frac{dy}{dx} = \frac{dy}{du} \times \frac{du}{dx}.$$

First we can rewrite the fraction as $(\pi x^3 + 3x^2 + 20x + 5)^{-4}$. Recall that if $f(t) = t^n$, then $f'(t) = nt^{n-1}$. So, obviously,

$$\frac{d}{dx}x^{-4} = -4x^{-5}.$$

However, we have an *expression* raised to the –4th power, not just an x. So we must use the chain rule. Using the chain rule, we have

$$\frac{d}{dx}u^{-4} = u'du = -4u^{-5} \times du.$$

Thus,

$$\frac{d}{dx}(\pi x^3 + 3x^2 + 20x + 5)^{-4} = -4(\pi x^3 + 3x^2 + 20x + 5)^{-5}$$

$$\times (3\pi x^2 + 6x + 20) = -\frac{12\pi x^2 + 24x + 80}{(\pi x^3 + 3x^2 + 20x + 5)^5}.$$

Don't you feel good now that you have solved this problem?

8. Rewrite

$$\frac{1}{\left(ex^3 + 2x^2 + 10x + 5\right)^5}$$

as

$$\left(ex^3 + 2x^2 + 10x + 5\right)^{-5}.$$

Use the chain rule to differentiate as follows:

$$\left[\left(ex^3 + 2x^2 + 10x + 5\right)^{-5}\right]' = -5(ex^3 + 2x^2 + 10x + 5)^{-6}(3ex^2 + 4x + 10)$$

$$= \frac{-5(3ex^2 + 4x + 10)}{(ex^3 + 2x^2 + 10x + 5)^6}.$$

9. We are asked to find

$$\int \frac{e}{\sqrt{x - \pi}} dx.$$

One way to solve this is to let $u = x - \pi$. Then $du = dx$. Notice that e is just a constant:

$$\int \frac{e}{\sqrt{x - \pi}} dx = \int \frac{e}{\sqrt{u}} du = e \int u^{-\frac{1}{2}} du = 2eu^{\frac{1}{2}} + C$$

$$= 2e\sqrt{u} + C = 2e\sqrt{x - \pi} + C.$$

10. If

$$a = \frac{d}{dx}\left(\tfrac{1}{e}x + \pi e^3\right),$$

then $a = 1/e$. Let $u = x - \pi$ so that $du = dx$. Integrating the first integral, we have

$$\int \frac{e^{-1}}{\pi\sqrt{x-\pi}}\,dx = \frac{1}{e\pi}\int\frac{1}{(x-\pi)^{\frac{1}{2}}}\,dx = \frac{1}{e\pi}\int u^{-\frac{1}{2}}\,du$$

$$= \frac{1}{e\pi}\times 2u^{\frac{1}{2}} = \frac{2\sqrt{x-\pi}}{e\pi} + C.$$

Integrate the second integral, and we have

$$\int 2x^3\,dx = \frac{2}{4}x^4 + C = \frac{1}{2}x^4 + C.$$

Therefore, our solution is

$$\frac{2\sqrt{x-\pi}}{e\pi} + \frac{1}{2}x^4 + C.$$

Further Reading

You're drifting into deep water, and there are things swimming around in the undertow you can't even conceive of.

— Stephen King, *Insomnia*

Colin Adams, Abigail Thompson, and Joel Hass. *How to Ace Calculus: The Streetwise Guide* (New York: Freeman, 1998).

———. *How to Ace the Rest of Calculus: The Streetwise Guide* (New York: Freeman, 2001).

Sheldon Gordon. *Calculus and the Computer* (Boston: Prindle, Weber & Schmidt, 1986).

Jan Gullberg. *Mathematics: From the Birth of Numbers* (New York: W. W. Norton, 1997).

Elliot Mendelson. *3000 Solved Problems in Calculus* (New York: McGraw-Hill, 1988).

Frank Morgan. *Calculus Lite*, 2nd edition (Wellesley, Mass.: A. K. Peters, 1997).

Clifford Pickover. *Wonders of Numbers* (New York: Oxford University Press, 2000).

Index